The Driftwood Legacy

A GREAT USIN' HORSE AND SIRE OF USIN' HORSES

by Phil Livingston & Jim Morris

featuring illustrations by Katy Peake

*Dedicated to Driftwood,
the bay stallion whose progeny has performed so well
for the men and women who make their living a-horseback.*

~

*Also dedicated to Katy Peake
who loved Driftwood and believed in his unique ability
to sire a superior strain of performance horses. Because of her efforts,
Driftwood still lives on in the veins of countless good usin' horses.*

Our appreciation to the late Catherine Peake Webster who made available her mother's collection of photos, breeding, and sales records, plus her own memories of the people, the horses, and the events of her Rancho Jabali years. Without her contributions, this book could not have been written.

Photo by Willard H. Porter

Driftwood and Katy Peake. *If it hadn't been for her dedication Driftwood might have been just another forgotten stallion in the history of the Quarter Horse. Katy Peake was a great horsewoman who believed in Driftwood's ability to sire great usin' horses.*

THE DRIFTWOOD LEGACY

Copyright © 2002 & 2010
By Phil Livingston & Jim Morris
Published By Wild Horse Press
An Imprint of Wild Horse Media Group
P.O. Box 331779
Fort Worth, Texas 76163
1-817-344-7036
www.WildHorsePress.com
ALL RIGHTS RESERVED
1 2 3 4 5 6 7 8 9
Paperback ISBN 978-1-68179-231-6
Hardback ISBN 978-1-68179-265-1
ebook ISBN 978-1-68179-271-2

Design, Typography and Production:
Kate Hamilton ~ Iron Horse Studio
Ridgway, Colorado

The authors and publisher shall have no liability nor responsibility with respect to any loss or damage caused or alleged to be caused directly or indirectly by the information or photos contained in this book. While the editorial and graphic content of this book is as accurate as the authors can determine from careful research, interpretation is dependent upon the reader's perspective.

While the authors have made every effort to insure that the Driftwood story has been completely told it is possible that incidents and individuals have been unknown or overlooked. For that, the authors apologize. In some cases the individuals and horses appearing in photographs are not known and those individuals not contacted.

The authors wish to thank the contributors of the rare photographs and Katy Peake sketches appearing in this book. Without them, only a part of the Driftwood story could have been told.

ACKNOWLEDGEMENTS

This book documenting the Driftwood legacy could not have been compiled without the help of countless friends, horsemen, and ropers who took time to remember their years with the Driftwoods and pass that information on to the authors.

A special thanks to:
Perry Cotton, Pleasanton, California
Jake Kittle, Patagonia, Arizona
Freeland Thorson, Nampa, Idaho
Mel Potter, Marana, Arizona
Gene Moench, Valparaiso, Indiana
Henry Kibler, Chandler, Arizona
Amy Turner, Denton, Texas
Johnny Burson, Silverton, Texas
Tom Eliason, Gregory, South Dakota
Babbitt Ranches, Flagstaff, Arizona
John Fincher, Plattsmouth, Nebraska
Kenny Nichols, Waco, Texas
John Balkenbush, Conrad, Montana
Buck Nichols, Gilbert, Arizona
Will & Cheryl Hall, Paradise Valley, Nevada
Sammy Fancher Brackenberry, Fort Tejon, California
Will Gill Jr. & David Gill, Madera, California
Jim Morris, Exeter, California
Ed Roberts, Fort Worth, Texas
Brian Wright, Arlington Heights, Illinois
and
other dedicated Driftwood owners, breeders, and horsemen.

TABLE OF CONTENTS

1 ~ The Driftwood Legacy 1

2 ~ The Early Years 5

3 ~ The Peakes and Driftwood 21

4 ~ Rancho Jabali 35

5 ~ Other Stallions at Rancho Jabali 47

6 ~ The Rancho Jabali Mares 51

7 ~ Driftwood Get Who Made Their Mark 61

8 ~Driftwoods and the Toughs Who Rode Them 79

9 ~ The Line Continues 87

10 ~ Keeping the Flame Alive 101

11 ~ A New Century 125

Appendix 1 ~ AQHA Get of Sire Summary for Driftwood .131

Appendix 2 ~ AQHA Get of Sire List for Driftwood 133

Appendix 3 ~ The Produce of Driftwood Daughters . . .137

Bibliography 145

1

THE DRIFTWOOD LEGACY

"A slender-barreled, deep rich bay who stood about 15 hands and weighed just ten hundred and seventy-five pounds in working shape."

—Katy Peake

1932	Driftwood foaled Silverton, Texas
1934	Sam & Amos Turner purchased Driftwood Silverton, Texas
1936	Amos Turner moved Jacksboro, Texas then Stony, Texas
1939	Buck Nichols & Ross Brinson purchased Driftwood Higley, Arizona
1940	George Cline family purchased Driftwood Tonto Basin, Arizona
1941	Asbury Schell purchased Driftwood Tempe, Arizona
1943	Channing & Katy Peake purchased Driftwood Lompoc, California
1946	Driftwood AQHA registration
1960	Driftwood's death
1983	Driftwood inducted into Trail of Great Cow Ponies, Cowboy Hall of Fame
1998	Katy Peake's death

The reputation of an aged stallion, and his ability to pass on the athletic talents which made him famous during his lifetime, seldom lasts more than two or three equine generations. Then, as those traits wear thin in his descendants and more fashionable bloodlines come to the fore, the horse is forgotten. It is only if the individual is so prepotent that his sons and daughters are able to perpetuate his abilities that breeders continue the line—that a stallion becomes immortal. Driftwood (known as Speedie during his rodeoing days) was such a horse.

That reputation is also dependent upon who owned a certain stallion. To Driftwood's good fortune, he belonged to horsemen and women throughout his life who appreciated his potential. First was Amos Turner, who raced and bred him. Ross Brinson saw and recognized the stallion's potential at an early age and urged his purchase by Nichols. Then, it was Buck Nichols, learning of the horse and making the long trip from Arizona to Texas to own and train him as a rope horse. Nichols was followed by the Cline family of the Tonto Basin in Arizona, who also roped on the speedy bay. Next was Asbury Schell of Tempe, Arizona, who gave the stallion national exposure as a rodeo mount. Lastly, there was Katy Peake, of Lompoc, California, who owned Driftwood the last seventeen years of his life. She deeply loved the horse, believed in him as a sire, and made every effort to obtain good mares which would help the stallion perpetuate his exceptional athletic ability for generations to come. Without her dedication, Driftwood would have probably slipped into the void with other forgotten stallions after his performance years and career as a stallion were over. To Katy Peake must go the credit for the continuance of the Driftwoods for half a century.

Driftwood was a winning race horse and an outstanding rodeo mount. That is enough to have gained him a brief reputation among horsemen. More importantly, he passed on his speed, athletic talents, his functional good looks and willing mind to his get. They, in turn, carried those same traits on to their offspring—

a legacy which has continued for more than fifty years. Only a few stallions have been so prepotent. With such a record, it's no wonder that usin' horse men have continued to breed the "Driftwoods" and ropers still cinch their saddles on them.

As a sire of rodeo horses during the 1940s, the 1950s, and into the 1960s, Driftwood was preeminent. As large a proportion of winning ropers went to the pay window on the capable backs of his sons and daughters as those of any other stallion. Even today, many of the top ropers and barrel racers are riding horses whose lineage can be traced back to the bay stallion. Among the stock horse competitors on the Pacific Coast during the middle years of the century, the athletic Driftwoods carved out a niche that is still remembered. Granted, those horses were ridden by some of the best horsemen of the period. They would not, however, have been in such demand if they could not do the job under saddle, stand up under heavy hauling and stay sound during their years of use.

Fig. 1-1. Photo by Willard H. Porter.

Driftwood, 1932 - 1960. *A head shot of Driftwood showing the calm, alert way in which he viewed the world.*

As an individual Driftwood was, in Katy Peake's words, "A slender-barreled, deep rich bay who stood about 15 hands and weighed just ten hundred and seventy-five pounds in working shape. He was a mighty little horse and in his lifetime did all the things asked of him and did them supremely well. He set a record for versatility and toughness difficult to equal and left behind a legacy of diverse accomplishments among his descendants which continues to proliferate. The presence of his name in a pedigree has become a hallmark of the qualities Quarter Horse breeders seek in establishing the highest standards of performance in their programs."

It couldn't be said better.

As with all horses of renown, there is the element of luck in Driftwood's career. Why, of all the fast horses available in his time, was he allowed to remain a stallion? Southwestern horsemen have always been quick to use the knife. There must have been something special about the bay stallion that kept his various owners from gelding him. Was it his speed, the fact that he had been bred at an early age and his colts were already proving out? Perhaps it was his disposition and the fact that he was "no trouble to have around?" Both as a race and rodeo horse, Driftwood had ample opportunity to create the problems that many stallions could and it didn't take much to condemn a troublemaker to life as a gelding. In any case, it was fortunate for succeeding generations of horsemen that he remained a stallion.

Another person who undoubtably contributed to Driftwood's notoriety was Willard H. Porter. As a freelance writer for horse publications, editor of *The Quarter Horse Journal* and publisher/editor

Driftwood, *by Miller Boy and out of a Comer mare. This picture shows his well-balanced, athletic good looks, smooth muscling and functional conformation.*

of *Hoofs and Horns,* Porter frequently wrote about Driftwood and the exploits of his get. In his many articles he always paid tribute to the good mounts the bay stallion had put under money-winning ropers. That constant exposure through the equine media had to attract attention to Driftwood and result in either breedings to the horse or sales of foals.

Even though Driftwood died in 1960, horses carrying his blood are still in demand. A few far-thinking breeders perpetuated the strain, carefully line breeding and then making sure that the resulting foals went to horsemen who would appreciate and use them. The roping world never forgot the Driftwoods. They continued to be sought after by men who valued a horse by his performance, functional conformation, athletic ability and willing mind. Not only was Driftwood roped on by champions but his descendants have been as well. A list of the outstanding ropers who have won on the Driftwoods is a "who's who" of the sport.

During the last decade, the Driftwood interest has blossomed. More and more horse lovers are rediscovering the performance qualities and the sturdy conformation of the Driftwood horses. Always popular with the cowboys, Driftwood enthusiasts have grown from the few scattered ranches which maintained the line as a source of working horses. Ropers, stock horse contestants, and folks who just like to ride a good usin' horse are paying premium prices for the foals carrying the blood of the bay stallion. The pendulum has swung back to horses that can do something under saddle. The Driftwoods ARE that kind of horse.

Driftwood's ability as a sire was recognized by horsemen from all areas. Frank Vessels, one of the great breeders of running Quarter Horses and the developer of the Los Alamitos Race Track, was visiting with Freeland Thorson, then Secretary of the Pacific Coast Quarter Horse Association. He made the comment, "If I had bought Driftwood the same day I bought Clabber from Ab Nichols, he could have made as great a broodmare sire as was Clabber." Vessels was referring to race horses, but he recognized the performance ability and speed of Driftwood's get.

Jimmy Williams of Rancho Santa Fe, California, was one of the finest hackamore and bridle horse men, in addition to being an outstanding trainer of jumping horses, during the 1940s, '50s and '60s. He trained and showed three Driftwoods, Woodwind (Hot Toddy), out of a Thoroughbred mare, Buttonwood, and Henny Penny Peake, out of Sage Hen, to multiple championships in both hackamore and bridle at Pacific Coast shows. He said of Driftwood's colts, "They are the best. You ask 'em to do anything and they'll do it. They want to learn and they have the ability to do something when you're finished with them. I think Speedy is as good a sire as there is any place."

Perry Cotton commented that "Driftwood was one of the greatest horses I ever knew. How many stallions have stood the test of time as he has—and the current revival of Driftwood interest is carrying on something which we knew fifty years ago."

This book is intended as a tribute to Driftwood, to the men who raced and rodeoed him, and to Katy Peake, who gave the bay stallion the opportunity to become one of the greatest performance sires in the history of the American Quarter Horse.

Fig. 1-3. Dardon Jasper photo.

Miller Boy and Milt Jasper (owner) holding his chaps. *This photo was taken at Silverton, Texas in 1922 when the horse was 5 or 6.*

2

THE EARLY YEARS

"He was a mighty little horse and in his lifetime did all the things asked of him and did them supremely well."

—Katy Peake

Like so many early-day Quarter Horses, Driftwood was a product of the Lone Star State. He was foaled near Silverton, above the Caprock where the plains drop off towards Quitaqui, Floydada, and Matador. The year was 1932 and his breeder was Mr. Bailey Childress (known in the area as "Old Man" Childress).

At that time, the Quarter Horse was not officially recognized as a breed. There were no official breeding records and pedigrees were loosely kept. Often, if they were racing their horses, owners preferred that the breeding be unknown. It was easier to match a horse if he was not known to be sired by speed.

The American Quarter Horse Association was formed in 1940 to document and record the breeding of early Quarter Horses. Driftwood was not included in the original registry. His breeding, like that of many of the early horses, was in the memories of the men and women who knew him. The major factor was his performance. The Peakes acquired the bay stallion because of his record as a race horse, rodeo horse, and sire, his functional conformation, and the assumption that he had solid Quarter Horse bloodlines. It wasn't until 1946—well after Katy and Channing Peake purchased Driftwood—that his history and breeding were traced and verified, and he was registered. The records carried by the American Quarter Horse Association list the bay stallion as being sired by Miller Boy and out of the Comer Mare.

Little is known about Miller Boy. He was said to be by the Hobart Horse. Again, there isn't much information about that individual other than that he was brought to the JA Ranch outside of Clarendon, Texas around 1920 by ranch manager T. D. Hobart. Old-time Texas horsemen, including Johnny Burson of Silverton, said that Miller Boy's dam, the Wylie Mare, was by Texas Chief by Lock's Rondo and out of a daughter of John Wilkens by Peter

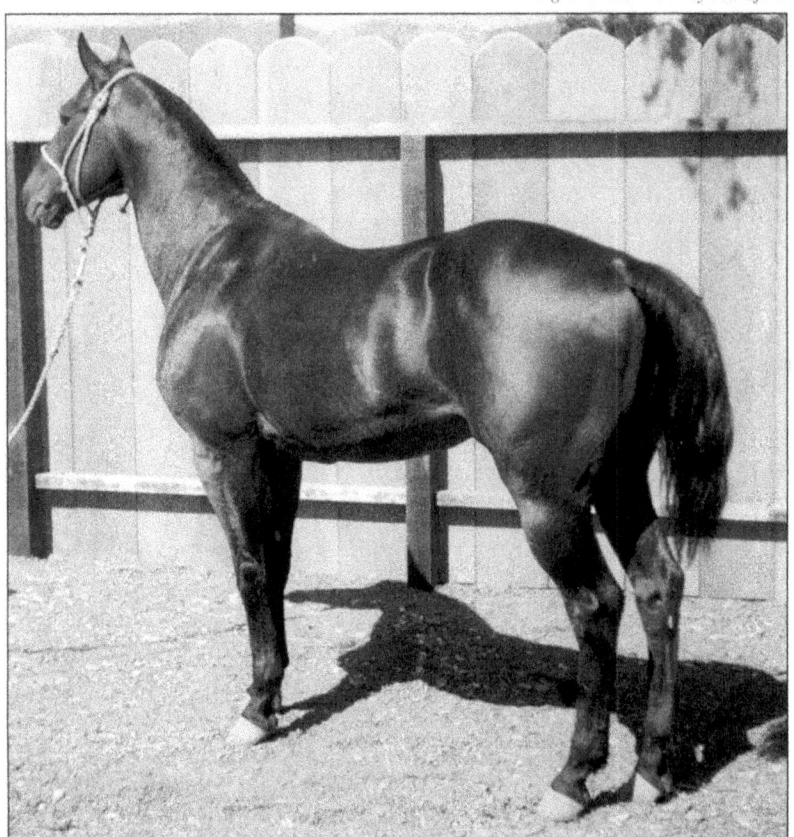

Fig. 2-1. Photo courtesy Peake files.

Driftwood "Speedie" in 1942 *as a 10-year-old showing the overall balance, the functional conformation and alertness that characterized him and his get. Photo taken at Rancho Jabali during Schell's visit.*

McCue. If so, Driftwood would be related to Joe Hancock, who was sired by John Wilkens. Miller Boy was bred by Will Wylie of Palo Duro, Texas and later owned by Milt Jasper of Silverton.

The Comer Mare, owned by Mr. Bailey Childress, was sired by Barlow, also by Lock's Rondo, and out of a Kentucky Thoroughbred mare. Again, the story cannot be verified. Only the remembered comments by old-timers give any indication of her pedigree.

According to this information, Driftwood was a line-bred Lock's Rondo, a horse that had speed to spare, plus a dash of Peter McCue through John Wilkens, another speedster.

Another version of Driftwood's breeding appeared in print in the 1944 edition of *Racing Quarter Horses,* prepared for the Arizona Quarter Racing Association by Melville Haskell. The listing concerned a 1935 son of Driftwood, Cowboy Schell, who was making a name for himself on the short tracks during the late '30s and '40s. The pedigree of Cowboy Schell showed Driftwood's sire as Line Up (TB) (Upset/Perling), a U.S. Remount Service stallion standing at Silverton in 1931. The same information appears in the 1945 and 1946 editions. It is known (from Silverton natives) that "Old Man" Childress took mares to both Line Up (TB) and Miller Boy, and since no paperwork existed it was possible for confusion on the breeding of his horses to develop. Johnny Burson, whose father stood Line Up (TB) from 1931 through 1936, does not believe that Driftwood was by Line Up (TB).

"If Driftwood was by Line Up," commented Burson, "The mare was bred some night when we weren't around."

Many of the early horses registered with the AQHA have the same cloudy background.

Channing and Katy Peake were aware of the possible Thoroughbred ancestry, but had also heard that the bay stallion had been sired by Miller Boy. They spent several years attempting to trace down the true breeding. Only after much research and a visit to Silverton in 1946 (following World War II), to talk with people who remembered the stallion, was it decided that Driftwood was sired by Miller Boy, and he was so registered.

Today, no one knows for sure, nor does it really matter. The breeding of numerous early-day Quarter Horses is similarly confused. Driftwood's sensational speed and outstanding physical ability, plus the genes to pass on the same characteristics to his get, prove he had exceptional bloodlines even if they are not exactly known.

During the 1920s and '30s the proving ground for the Quarter Horse was the match tracks. There were no horse shows, professional rodeo was in its infancy, and there was no Quarter Horse Association to record bloodlines. A stallion who could run was in demand

Fig. 2-2. Photo courtesy Turner files.

Driftwood as a 2-year-old with Amos Turner. Location not known.

because his colts could probably also race successfully, catch a cow, or play polo. Since Driftwood's pedigree read speed regardless of his sire's breeding, he went to the race track. There were no recognized meets holding quarter mile races and the sport was confined to the brush tracks throughout the Southwest. Stallions such as Steeldust, Shiloh, Traveler, Peter McCue, Joe Hancock, Badger, John Wilkins, Midnight, Oklahoma Star, Possum (King), Gray Badger II, and Zantannon all made their reputations at those dusty meets before retiring to stud duty.

It is not recorded how long Mr. Childress owned Driftwood. Sam Turner, also of Silverton, acquired the horse, probably before he was a two-year-old. His son, Amos, liked the bay colt and purchased him. Under the handling of Amos Turner, Driftwood rapidly established himself as a horse to beat on the Panhandle tracks.

Wilma Turner, wife of the late Amos Turner, remembered the stallion well. "Driftwood was a real nice gentle horse and he was always ready to run. He was very fast and we never did find a horse that could beat him. We used to use a neighbor boy as our jockey."

Texas has always been fast horse country and it didn't take long for the word to get out that the Turners had one that could fly. That meant that challengers kept coming by. Rather than race against all comers, Amos Turner had a system to see if a horse would get the opportunity to try Driftwood.

"When we lived on the ranch (Silverton), my brother Shorty Wood lived with us," recollected Mrs. Turner. "He grew up with the horse and he had a paint horse he rode a lot. When people came out to see how fast a horse they had, they had to run the paint first. Of course, he could usually beat them. Mr. Turner wouldn't match Driftwood against the ones the paint horse beat because he was so much faster. He was so fast that all the horsemen wanted to know if they had one that could beat him."

Mrs. Turner didn't attend all of the races, but she did remember one in particular. "One time some of those guys thought they had horses that were pretty fast so they were going to cut Driftwood off and beat him. Well, the horses got tangled up and one of the boys was killed. I felt bad that I had a horse involved in that. But the mixup didn't bother Driftwood. He went on and won the race."

Johnny Burson, who grew up in Silverton, neighbored to the Turner ranch. The first time he remembered seeing Driftwood the two-year-old was staked out on a long rope tied to a log and being halter broken. He was ganted up and rope-burned. Turner had turned the colt over to Dewey Beavers and a man named Woods for breaking and training.

The next time Burson saw the horse, he was "fat and pretty" and being galloped on the track Turner had worked up in the pasture.

Asked if he knew where the name "Driftwood" came from, Burson commented that he didn't. "Most of the time, everyone just called the bay stallion the Amos Turner horse, a common way of referring to horses at the time."

Johnny recalled seeing Driftwood run—and win—when he was just a teenager. It was the first time he had ever bet on a horse race and he put $5 (a sizable bet during the early 1930s) on the bay. He collected $8 after Driftwood crossed the finish line first. The bay Driftwood was matched against a sorrel (breeding unknown) also called Driftwood.

Burson later watched him run 300 yards at Lubbock, Texas. It was the first paramutuel betting race (paramutuel was legal in Texas at the time) he had ever witnessed, and Driftwood was announced beforehand as being for sale. He doesn't remember what horse Driftwood was matched against, but the stallion was ten lengths ahead at the finish.

"Buck Nichols and another fellow were there to see him run and that's when he went to Arizona," remembered Johnny. "The men came over to the house the night before and talked about the horse. They also bought a sorrel three-year-old from us, a colt by Line Up (TB) and out of a Chickasaw Bob mare."

There were no starting gates on the brush tracks where Driftwood ran. The horses were brought up to the line and flagged, or "tapped off" by the starter if they "lapped" each other. If not, they were called back. Frequently, it took several tries for the horses to get an even start. Some horses couldn't take the pressure of repeated starts and then being pulled up. They "blew" or became so nervous that they couldn't run their best in the actual contest.

The term "tapped off" came from colonial days when the starter either tapped a stick against a board or had a man beat on a drum to signal the riders it was an official start.

Fig. 2-3. Photo courtesy Turner files. Reprinted from The Foundation Quarter Horse Journal.

DRIFTWOOD BREEDERS

Driftwood & Amos Turner.

Amos Turner on Driftwood — *time and location not known.*

Race horse trainers of the time would spend hours teaching their horses to come up to the line, to break and then pull up. It was called "scoring." Sometimes the practice went on for hours to see which horse would blow up first.

Willard Porter, in his book *13 Flat,* recounted a conversation with Katy Peake about Driftwood's ability to remain calm despite repeated false starts. An old Texas racehorse man told her that "Driftwood never blew. Once he was matched against one of the best lap-and-tap horses in the Sweetwater area for 220 yards. The horses scored for an hour and a quarter, until both animals were drenched with sweat from so many unlapped starts. The other horse was high, nervous and obviously tiring, but Driftwood was still calm. When they finally broke, he was on top and won by two lengths."

Johnny Burson was a fan of Driftwood and tried to get his father, "Long John" Burson, to buy the horse. To him, Driftwood would fit into the family program of owning fast horses. He didn't succeed since the elder Burson liked a bigger, stouter animal and Driftwood was "small boned and little." "Daddy felt that Driftwood's ankles and pasterns were too small and that he wouldn't stand up under hard use." The Burson ranch raised horses for many years, standing the Remount Thoroughbred stallions Battemout, Line Up and Reno Headlight as well as Texas Chief. Johnny Burson was to have an impact on the Driftwood story. He owned and stood Tater 3811, by Kiowa and out of Clayton Gal by Driftwood. Two Tater geldings that did "right well" in the rodeo arena were Heelfly, under Ike Rude, and Shortgrass, a good calf roping mount for T. C. Jones and Sonny Hendrix.

In 1936 (when Driftwood was a 4-year-old) Amos Turner moved from Silverton to Jacksboro in north-central Texas. Several months later, he moved again, to Stony, near Denton. He farmed, drove a truck and a school bus, and ran a threshing machine. Since he brought Driftwood as well as several mares and yearlings with him (the breeding of those mares is not remembered but it was assumed that they were part Thoroughbred), he stayed in the horse business. He continued to match his speedy bay stallion. When Turner couldn't find a horse race, he was willing to run Driftwood against an automobile for any distance up to a quarter of a mile. Country roads were dirt, and while the car was spinning its wheels, Driftwood was sprinting down the course. O. D. "Bud" Smith was Driftwood's regular jockey during those years.

Jack Brisco, now of Ponder, Texas, was a boy at the time and lived next door to Amos Turner. He spent lots of time on the Turner front porch listening to the men "talk horses." He remembered Driftwood as a horse that could run pretty well. Jack did see him race a few times, recalling that he was never headed. One incident at nearby Denton stayed in his mind, not so much for the race but for what happened before.

One weekend, as Brisco remembered it, Amos Turner loaded Driftwood into the back of his brand new 1937 Chevrolet pickup (with wooden sideboards to form a stock rack since horse trailers weren't in common use at the time) and headed for Denton. There was a "racin' and ropin'" scheduled and he was planning to see what kind of match he might scare up. When he got to the track, near the railroad tracks, the whistle of a passing freight train spooked the stallion and he climbed out—over the cab of the pickup.

While on the Stony farm, Turner sold the Driftwood colts and mares which he'd brought from Silverton. What became of the mares is not remembered. Three geldings became well known in the Denton area as top mounts. Smokey, owned by Chet January, was the kind of horse that "you could do it all on." They worked cattle during the week and roped calves, match raced and picked up broncs on him on weekends. George Seals, from nearby Ponder, remembered the horse well.

"Anybody could ride Smokey. If you could rope at all, he'd let you win," recollected Seals with a smile. "Chet hauled him around here in the back of a pickup with no sideboards. Old Smokey would jump in and out anywhere. We used to go get him when we gathered wild cattle and hauled him that way." Seals also recalled the time that a bull hooked the horse in the side. "They stitched him up and he got all right."

Since, like his sire, Smokey could run pretty well, he was occasionally matched. One time, at Harlee Field in Denton, some men got together and brought in an old nag with harness on him to race against Smokey. Unbeknown to Chet January, the other horse had received, in Seals' words, "a little help" and managed to beat Smokey across the finish line.

Goober, a full brother to Smokey, was another good roping and ranch horse. Bill Bentley, at nearby Sanger, owned Goober and he'd let almost anyone

Fig. 2-4. Photo courtesy George Seals.

Goober, *an unregistered son of Driftwood foaled before the formation of the American Quarter Horse Association, with* **George Seals** *roping on him at the Denton Creek Arena, Justin, Texas in the early 1950s. Goober was a ranch horse during the week and a rodeo mount on weekends.*

rope on the horse. "Red" and Billy Joe Deussen, George and Bobby Seals, Junior Beville, and a host of other Texas ropers all got a seat on the dependable Goober at the local rodeos.

Another Driftwood colt in the area was Snake. He was owned by Wylie Ennis, who also made a practice of letting other ropers ride him. Like his half brothers, Snake earned his oats as a ranch horse and rodeoed during weekends.

Even then, horsemen were aware of the versatility, working ability and good minds that the Driftwoods had.

George Seals remarked, "There was something in that Driftwood breeding. His colts were quiet, athletic and would do whatever you wanted them to do. There will never be another stallion like him."

Amos Turner soon discovered that he had more jobs than he had time for and got out of the horse business. Since he was mainly interested in farming, he traded Driftwood to his father for some work horses and sent the bay stallion back to Silverton.

The senior Turner sold the stallion. Silverton native Ross Brinson was involved in the transaction. Brinson's family had moved to Higley, Arizona, where he went to high school with Buck Nichols. On a trip back to Texas, he visited Silverton and found out that Driftwood was for sale. Brinson returned to Arizona, told Buck Nichols about Driftwood, and the two men worked out a partnership. According to Dave Stout, former editor of *The Rodeo Sports News* and later Secretary of the Rodeo Cowboys' Association, the deal was that the two men would buy the horse together, if Buck could "put a stop and get-back on him" for a calf roping mount and would split the profit when he was sold. The two men (probably in 1939) went to Silverton to look at Driftwood. (Kenny Nichols remembers his grandmother talking about being left behind to take care of the children while her husband went to Texas.) Buck Nichols liked what he saw and the deal was made—adding another link to the legend. Grandson Kenny also said that Brinson purchased another horse from Johnny Burson. That animal, thought to be a half-brother to Driftwood, was sold during an overnight stop in New Mexico.

Buck Nichols, the seventh of ten children, was born in Cheyenne, Oklahoma (stomping ground of Peter McCue for awhile) on August 4, 1913 to Ab and Mabel Nichols. The family moved to Arizona in 1918 where they developed a substantial farming and ranching operation. They also raised fast horses and Ab was always willing to match a race. Among the speedsters the Nichols owned were Captain White Sox, Clabber (1940-41 Champion Quarter Running Horse), Colonel Clyde, Lucky, Lady Lux N, Wagon N, Woodfern, Frosty Joe, Blue Bonnet, and others. Buck grew up cowboying and training horses, and was a tough PRCA calf and team roper during the 1940s and '50s. He had the "horseman's eye" when it came to conformation and being able to see the potential in a prospect. During his long career, he bred and raised many outstanding individuals whose blood runs in the veins of today's top performance mounts. In Driftwood,

Fig. 2-5. Photo courtesy AQHA.

Clabber 507. *This 1936 son of Texas Dandy and Blondie S was owned by Ab Nichols at the time he purchased Driftwood. Since both Clabber (1940-41 World Champion Quarter Running Horse) and Driftwood could fly down the track, Nichols had to have one more race. Driftwood won.*

he saw the proven speed, conformation and willing disposition that he was looking for, and bought him immediately.

It was during his stay with Nichols that Driftwood, as the story goes, outran Clabber. The family owned both stallions and "had to see which one could get down the track the fastest."

Buck Nichols was quoted by Willard Porter in a *Hoofs and Horns* article as saying, "Driftwood was six or seven years old when we got him. He was a race horse but I started roping on him. As far as I know, he'd never done that before we got him."

Nichols continued, "He was a blood bay, not real big, and definitely showed his Thoroughbred breeding. And he could run! He was a smooth riding horse with an easy disposition."

Chilina, a 1950 Driftwood/Hancock Belle daughter, shows the Driftwood ability at roping with owner **John Fincher** at a 1950s jackpot at Sierra Vista, Arizona. Chilina was bred by Buck Nichols.

"We didn't keep him more than a couple of years, but we thought a great deal of him. As a matter of fact, we bred a number of our Clabber mares to him."

Since that happened before the formation of the American Quarter Horse Association, those foals were not registered and cannot be traced. It is safe to say, however, that the Driftwood/Clabber combination put a number of Arizona cowboys "a-horseback" and the fillies contributed to the quality of ranch remudas.

Even after selling Driftwood, Buck Nichols continued the bloodline in his horses. He not only kept mares at Rancho Jabali, picking up the foals when they were weaned, but trailered other mares to California for breeding. Driftwoods which he owned included Speedy II, Chilina, Chakaty and Woodfern (the dam of Oui Oui, who produced Wilywood when bred to Orphan Drift).

It was shortly after the Nichols acquired Driftwood that another man stepped into the picture. Roy Wales, a rancher, farmer, horseman and roper from nearby Queen Creek, was invited over to see Buck Nichols' new calf horse (Driftwood). He roped four calves on him and asked, "Buck, what are you riding at Prescott (a big Arizona rodeo held over the 4th of July)?" Nichols answered, "The bay." Roy quickly came back with, "We'll both ride him."

At Prescott, Wales broke the barrier on three of the four calves (carrying a 10-second fine each time) or he would have been fast enough to have won all four go-rounds and the average. Driftwood started hard and was running wide open the first stride out of the box. Roy stated that he let each calf "way out down the arena but still broke out." From that day on, almost every horse that carried Roy Wales or his family was a Driftwood.

Roy Wales once told Willard H. Porter that "When Buck Nichols first took Driftwood to those Arizona rodeos, none of us had ever seen a horse like him. There was just nothing like him in the whole state."

Wales bred mares to Driftwood as long as he was in Arizona and then followed the stallion to California. He even left mares at Rancho Jabali

Fig. 2-7. Photo courtesy Leckie Lobapesky.

Leck Cline *roping on* **Cowboy C** *at Prescott, Arizona in 1948.*

permanently, picking up the colts when they were weaned. He would halter break and turn them out on a half-section of desert pasture. The next year, the colts would be brought in, played with and then turned back out. When they were two, they were started under saddle.

During his long career with the Driftwoods, Roy Wales not only bred to the stallion, raised and campaigned his colts, but also stood a son, Driftwood Ike. The smokey dun stallion carried on his sire's reputation of not only being a good roping horse himself but sired horses that ropers could "win on." Today many of the top timed event cowboys are riding his descendants. Wales also owned and roped on three other Driftwoods of note— Firewood, Hallie Wood, and Wooden Nugget.

The Porter article continued with Buck Nichols' memories. "George Cline, of Roosevelt, Arizona, saw the horse and wanted to buy him. We sold him for $600—an unheard of price since the Depression of the 1930s wasn't over yet. In addition to ranching, the Clines were tough ropers and Driftwood got lots of arena experience." Proof of George and Leck Cline's ability with a catch rope was the fact that during their careers they won or placed at every big rodeo in the Sunshine State.

One old time roper remarked that "We were all glad that the Clines stayed home to ranch. They could rope and were always well mounted."

In addition to ranching and rodeoing, the Cline family also raised and raced horses on the Arizona tracks. There was even a race track and starting

Fig. 2-8. Photo courtesy Turner files.

Cowboy *(registered as Cowboy Schell 16612), foaled in 1935 by Driftwood and out of a Will Steed mare. Cowboy earned a ROM in Racing before becoming an outstanding rodeo mount for Abe Graham, Maynard Gaylor, Cliff Watley, Asbury Schell and Eddie Schell. This photo was taken when he was racing in Arizona.*

gate on the ranch (George's niece, Leckie Lobapesky, remembers playing on it when she was a child) where they trained their runners. One of the speedsters that the Cline family bred and owned was Prissy by Colonel Clyde, who set a world record for 350 yards and outran such greats as Miss Bank, Blondie; Queenie; and Buster ROM, by Clabber.

While the bay stallion was purchased primarily as a rodeo mount, there is no doubt that he was bred to a number of mares during his tenure at the Roosevelt ranch. One Driftwood foal, a 1940 stallion named Cowboy C, went on to become a top calf and team roping mount for the family, often going to a set of heels behind Asbury Schell riding Cowboy Schell. Cowboy C was listed as being bred by Joe Bassett but owned by George Cline. He was frequently ridden by brother Leck as well. A Cline-bred mare, Prissy Cline by Driftwood Ike, was the maternal granddam of French Flash Hawk "Bozo," three-time AQHA Barrel Racing Horse of the Year and the mount which carried Kristie Peterson to two WPRA Barrel Racing titles.

June Winters of Tonto Basin, Arizona, George Cline's daughter, remembers when Driftwood was at the ranch. "He was a wonderful horse, with a sweet, well-mannered disposition. He was bred to a few mares, but mostly Dad just used him—both as a ranch horse and a rodeo mount."

George Cline didn't own the horse much longer than a year before another man decided that he had to cinch his saddle on the good looking, good acting bay stallion. Driftwood was nine years old at the time.

Asbury Schell of Tempe, Arizona was a World Champion Team Roper and a tough calf roper. He had seen Driftwood at rodeos under Nichols, Wales and the Clines, and decided that he needed to own him. With Asbury Schell in the saddle, it wasn't long before the bay stallion was well known on the big-time rodeo circuit. He carried numerous riders from Salinas, California to New York City and Boston; from Douglas, Arizona to Calgary, Alberta. Because of the speed with which he caught cattle, Schell began calling the stallion "Speedie" (the Peakes later spelled it "Speedy") and the name caught on among the cowboys.

June Winters commented, "Dad and Asbury were good friends, in addition to being competitors in the arena. When Asbury wanted the horse, it was probably hard for Dad to turn him down."

During the 1930s, '40s, and '50s the majority of rodeos were held in large outdoor arenas and the roping scores (the start that the cattle are given from the chute) were long—sometimes up to 60

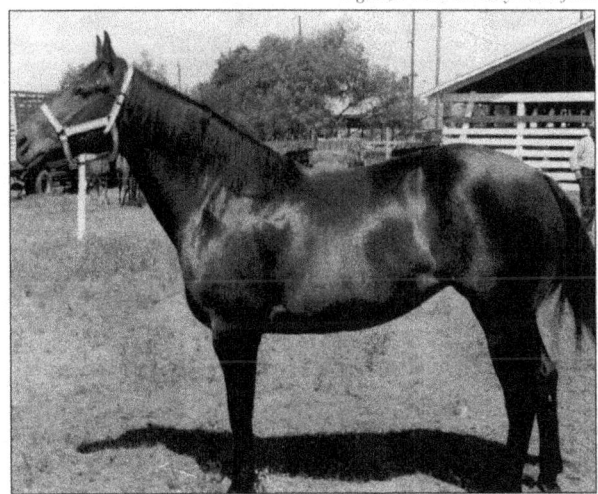

Fig. 2-9. Photo courtesy Peake files.

Wanda Marie 3250, by Driftwood and out of a Miller Boy daughter. She was foaled in 1939. As a broodmare she produced Tan Jug; Rancho Earl 2, Open ROM; Senior Driftwood (by Senior George), Open ROM, 1966 High Point Steer Roping Horse and Steer Roping Stallion; Troubadour Dee; Chubby Marie; show winner Goldie Binford; Midnight Mist; Senior's Nina; and Chester Jr.

Fig. 2-10. Photo courtesy Peake files.

Tater 3811, by Kiowa 1087 and out of Clayton Gal 3367 by Driftwood. He was bred and owned by Johnny Burson of Silverton, Texas. Tater was a top rodeo and ranch horse in addition to being a good sire.

feet. It took a fast mount to put his rider in roping position quick enough to win money. As a match race horse, Driftwood had learned to break hard and fast and was able to catch cattle in a hurry. And he would really hunt and track cattle. The horse also developed a "killing stop," one which would take the fight out of big, rank calves. Schell "did it all" on the stallion, roping calves, heading, heeling and single steer roping, as well as hazing bulldogging steers.

Old-time rodeo hands still talk about Driftwood's performance at the Payson, Arizona rodeo (probably 1941). The horse was ridden in every timed event—calf roping, team tying, single steer roping and bulldogging—and his riders collected checks in all of them. Then, just to show that he hadn't lost his quick speed, Driftwood won the cow pony race down the arena.

James Kenny, now of Midland, Texas, was rodeoing during the time that Asbury Schell was competing on Driftwood. He remembered that "The horse could sure catch cattle in a hurry."

There is ample evidence that Driftwood was used as a stallion at an early age. Cowboy (later registered under the name Cowboy Schell), was foaled in Texas in 1935. That meant that Driftwood would have been breeding mares as a two-year-old. Other offspring credited to him include Clayton Gal, foaled in 1936 (dam of Tater 3811); Wanda Marie, foaled in 1939; Sally Wood; Dixie Bell; and Dixie Bell National. Both George Cline and Asbury Schell bred him to a number of mares, but the majority of these foals were not registered. How many foals he sired, both in Texas and Arizona, cannot be verified.

The Driftwood legend was beginning.

In early 1943, Driftwood made the most important move of his life. He was purchased by Channing and Catherine "Katy" Peake, of Lompoc, California. Several years earlier the couple had established a Quarter Horse operation specializing in the production of ranch, rodeo and performance horses. The Peakes began with five RO mares from the Greene Cattle Company of Patagonia, Arizona. To these, they added another five mares of One Eyed Waggoner breeding, purchased from Duaine Hughes at San Angelo, Texas.

Fig. 2-11. Photo courtesy Western Livestock Journal.

Gordon Davis *of Templeton, California. This roper, rancher, and horseman is the individual who told Katy Peake about Driftwood.*

Fig. 2-12. Photo courtesy Peake files.

Driftwood with Asbury Schell *in the saddle, at Rancho Jabali in 1942. Channing and Katy Peake had seen the horse at the Hayward, California rodeo and asked Schell to come to the ranch and breed Driftwood to several of their mares.*

Fig. 2-13. Photo courtesy Foundation Quarter Horse Magazine.

*One of the few remaining photos of **Driftwood** during his rodeo career. Asbury Schell on **"Speedy"** coming in to the heels for Gordon Davis on the Paint at a Team Tieing.*

While the Peakes had the broodmares necessary to achieve their goal, they didn't have a stallion. There were numerous Quarter Horse stallions available at the time but none met the criteria the couple had laid down. They wanted a horse that had shown he could get the job done in the arena, stay sound, have speed and a good disposition. Conformation was to be dictated by the old adage, "Form Follows Function," not the current "showring style." And they wanted a stallion of proven prepotency, one which passed on those traits to his offspring.

They had been searching for some time. Then, one day, Gordon Davis of Templeton, California stopped by. He was a rancher and a roper who could compete with the best. According to Willard Porter, Davis told Catherine Peake, "Katy, I've got the horse for you. They call him Speedie and he's owned and ridden by Asbury Schell of Tempe, Arizona. He's a rope horse all the way!"

In the spring of 1942 the Peakes joined Davis at the Hayward, California Rowell Ranch Rodeo where Schell would be competing. They could not only see what the stallion looked like and assess his ability but talk with his owner as well.

Driftwood was everything that Davis had told them. He was not only good looking, with a quiet eye and kind disposition, but showed the rope horse quality the Peakes had been looking for.

There was only one problem. Asbury Schell didn't want to sell Driftwood. The horse "fit" him, something that ropers treasure in a mount. Channing and Katy Peake tried hard but weren't able to change the cowboy's mind. They did, however, talk him into visiting their ranch and breeding Speedie to several mares.

When Asbury loaded up to return to Arizona, the Peakes asked for the first option to buy the horse if he ever decided to sell him. Schell agreed—although he didn't give them much encouragement.

After he returned to Arizona, Asbury Schell decided that he would part with Driftwood. World War II had begun and rodeo travel would be curtailed. The cowboy really wasn't sure of what he wanted to do. Yes, he was winning off of the horse, he enjoyed owning him—but he had to be able to get to the rodeos to compete. National gas rationing (a part of the war effort) would make that difficult. He wrote Chan and Katy, informing them of his desire to sell Driftwood. The Peakes agreed to purchase the stallion.

Then, Schell changed his mind. Correspondence went back and forth between Tempe and Lompoc as he worried over the situation. The following letters, primarily from Asbury Schell to the Peakes, show the feelings of both parties. Schell's indecision and the persistence with which the Peakes pursued their goal are evident.

October 21, 1942
Mr. & Mrs. Channing Peake,
Dear Friends:

How are you folks by this time? Fine I hope. This leaves us all OK except we have been mourning over Speedie.

Say Chann I suppose you will think I am a cheap skate, but I am asking you to let me back out on the deal. I would have to bury the whole family if I let him go, including myself.

If you ever want to use him you are perfectly welcome to come and get him anytime. I hope this hasn't caused you any inconvenience, if it has, I will try and make it right. He is in the best shape I have ever seen him in and is sure working good.

Joe Bassett has offered to run his Speedie colt at anything in the state under three years of age.

I am hoping this doesn't cause any friction between us, if it is I had rather go ahead with the deal.

*Best Regards to all
Asbury Schell*

~

Fig. 2-14. Photo courtesy Peake files.

Joe Bassett's Speedie colt to which Asbury Schell referred in his letter. At the time the photo was taken, Bassett had offered to run the colt against any horse in Arizona under 3 years old. This horse is probably **Cowboy C** 8502, foaled in 1940.

October 24, 1942

Dear Asbury,

Katy and I have been planning for sometime to drive down to Phoenix this next week to see you and pick up Speedy. We figured that you would be back from the big roping and we want to get down there before gas is rationed.

Your letter came yesterday and we have been thinking over what you said in it. Before we come down I would like you to know how we feel and what we think about it. I hope you will be able to see things from our point of view. We are in the Quarter Horse business and we are trying to breed the best Quarter Horses possible. We have spent a number of years and a lot of money assembling a group of young broodmares. We know that you think they are good ones and ought to be mothers of some wonderful colts. We have to sell these colts to stay in the game and having a good, even outstanding bunch of mares will never sell colts to the public. Speedy, outside of being a horse of ideal type and breeding to cross on our mares, has a name that is known all over as a performer. That is what will sell colts to the people we want to sell to and it is the only yardstick you can have of a colt's future ability. We were looking for a stallion when you appeared like a lucky star on our horizon. We were more than delighted to be able to breed our mares to a horse of Speedy's caliber and more than ever realized that we would have to acquire an outstanding horse to give our ranch the kind of reputation we want it to have. Since you wrote about wanting to sell Speedy and we agreed to buy him, we have naturally not looked further for a stallion. We have more or less spread the word around that we had bought him and I am sure that Gordon Davis has done the same. It would cause us considerable embarrassment to be unable to produce him.

All this foregoing is naturally from our standpoint alone. We have tried to see the thing from your point of view. We know your personal attachment for Speedy. That is a point which bears no argument. However looking into the future it is more than apparent that the rodeo game is temporarily washed up, for how long none of us can judge but it will not be as short as we'd like to have it from any point of view.

Speedy is not a colt and his value as a rope horse will naturally diminish. He is a great individual and even from your outlook the part that he can play to the best advantage is the siring of colts out of mothers who have the best chance of producing his like. I don't know where he could be better off with that in mind than with us and I know that you agree with me there. What it really comes down to I guess is just not being able to part with him. We are still planning to see you around the middle of next week and in case we don't have to return with an empty trailer, please have a dourine test made on Speedy if you haven't already done so. We are looking forward to seeing all of you again.

*Sincerely,
Channing & Katy Peake*

~

October 27, 1942

Dear Chann,

Received your letter, was sorry of your attitude about Speedie. Speedie is not for sale. You see circumstances alters lots of things, as was the case when you bought Gordon's paint mare.

You see, Speedie has never been in my name. We are leaving for Coolidge Thursday. I am going to work down there.

I am sorry about this and was trying my best to be nice about it, but if that's the way you feel about it there is nothing more I can do. I really meant it when I said you can use him anytime.

If you really want a stallion, that horse of Joe Bassett's is as nice a horse as I have seen any place. Anyone that has seen him will tell you so. He only asks $1000 for him. I know that if you would see him you would say so yourself.

Regards to all,
Asbury Schell

~

On October 28, 1942 Asbury Schell confirmed his decision to keep Driftwood by telegraphing Channing Peake. The message was short—and definite.

```
WUX TEMPE, ARIZ   AM OCT 28, 1942

CHANNING PEAKE
RTE 1 BOX 170,
LOMPOC CALIF.

DONT COME AFTER HORSE.

ASBURY
1205PM.
```

Even in the face of a strong "NO" to their offer, Channing and Katy Peake kept the line of communication open and maintained an interest in purchasing Driftwood. Evidently, they realized that Schell wasn't really sure of what he wanted to do and they were determined that the decision he finally made would be in their favor. A letter early in the following year was evidence that they had read the situation correctly.

January 7, 1943

Dear Chann & family,

Received your most welcome letter several days ago, also your picture of myself and Speedie. It was sure good of him. We was sure glad to get it, and thank you very much for it.

We was sure glad you feel the way you do about everything. We sure felt bad about it, after you treated us so well last summer.

Maybe after the Phoenix rodeo I may change my mind about Speedie. If I do you will sure be the one to get him, if you still want him. I am not going to say anything about it until I am sure I will sell him.

We was glad to hear you got the place next door to you. I suppose you are quite busy now. I didn't take the job in Coolidge as things didn't turn out as I expected. I am a testing tanker for Uncle Sam about 15 miles from Tempe but we are living in the same place. If you happen over here, come and see us.

Hoping you folks had a Merry Christmas and a Happy New Year.

Respectfully,
Asbury Schell

~

January 24, 1943

Mr. Channing Peake
Dear Friend:

I suppose you will be surprised to hear from me so soon, but here goes again.

I have decided to let you have Speedie, just as soon as the Phoenix rodeo is over. It is March 5, 6 & 7th. If you care to you can send me a forfeit, and then I can't back out.

I don't want anybody else to have him but you folks. I wouldn't sell him to anybody that follows rodeos.

He is just as fat and sound as he can be. I will guarantee no backfire on this.

I am still on the job out here. I am sure anxious for my colts to arrive out of him.

I wish you folks could come over for the rodeo and look around a little.

Will say good bye for this time, hoping this finds all you folks well.

Your friend,

Asbury Schell
P.S. let me hear from you soon.

~

Needless to say, the Peakes promptly mailed an acceptance, an agreement of sale, and a forfeit check for $500 to Schell. Their continued efforts to purchase Driftwood had finally paid off. They also made plans to attend the Phoenix rodeo in March and take possession of the horse.

February 15, 1943
Friend Chann,

Rec'd your letter, check and papers, was glad to hear from you. I am signing the paper and returning it to you.

I am not kidding you when I say you are the only person I would let have Speedie. I would like to know what you and Katy would charge me for a good horse colt from him.

I would like to have a young Speedie of my own, and see if I can't make another Speedie of him. We will be looking for you at the rodeo.

your friend,
Asbury Schell

~

Channing and Katy Peake made the trip to Phoenix in March of 1943. Even though he had won more on Speedie at the rodeo than the purchase price, Asbury Schell delivered the bay stallion as agreed. After the final rodeo performance the Peakes handed over the balance of the purchase price, took possession of Driftwood and returned to Rancho Jabali.

A public announcement of Driftwood's (Speedy's) acquisition was made in the form of a half-page advertisement in the May 15, 1943 *Western Livestock Journal*. In October, a full page ad picturing Driftwood (Speedy), plus three mares and their first foals, appeared in the same publication. Showing photos of the foals was a sampling of what Rancho Jabali would be offering for sale in the future. It is interesting to note that, in the early

---

February 8, 1943

AGREEMENT OF SALE:
Asbury Schell to Channing Peake

Asbury Schell agrees to sell to Channing Peake one bay Quarter Horse stallion "Speedie", as understood by both parties, for the sum of fifteen hundred dollars ($1,500.00) and hereby acknowledges the receipt of a check dated February 8, 1943 for five hundred dollars ($500.00) from Channing Peake as a partial payment for the above mentioned horse.

It is understood that the balance of one thousand dollars ($1,000.00) will be paid upon delivery of a certified bill of sale for the horse to Channing Peake. It is further understood that the agreement shall be binding upon both parties and that the transaction shall be completed within sixty days (60) of this date (Feb. 8, 1943) unless the horse becomes unsound before Channing Peake makes the second payment, in which case Asbury Schell agrees to return Channing Peake the five hundred dollars ($500.00) received from him and this colt contract becomes null and void.

It is also understood that the term "unsound" shall be construed to mean defective in such a manner as to be unacceptable to the buyer.

Sale Agreement for Speedie. *This was signed by Channing Peake, and later by Asbury Schell.*

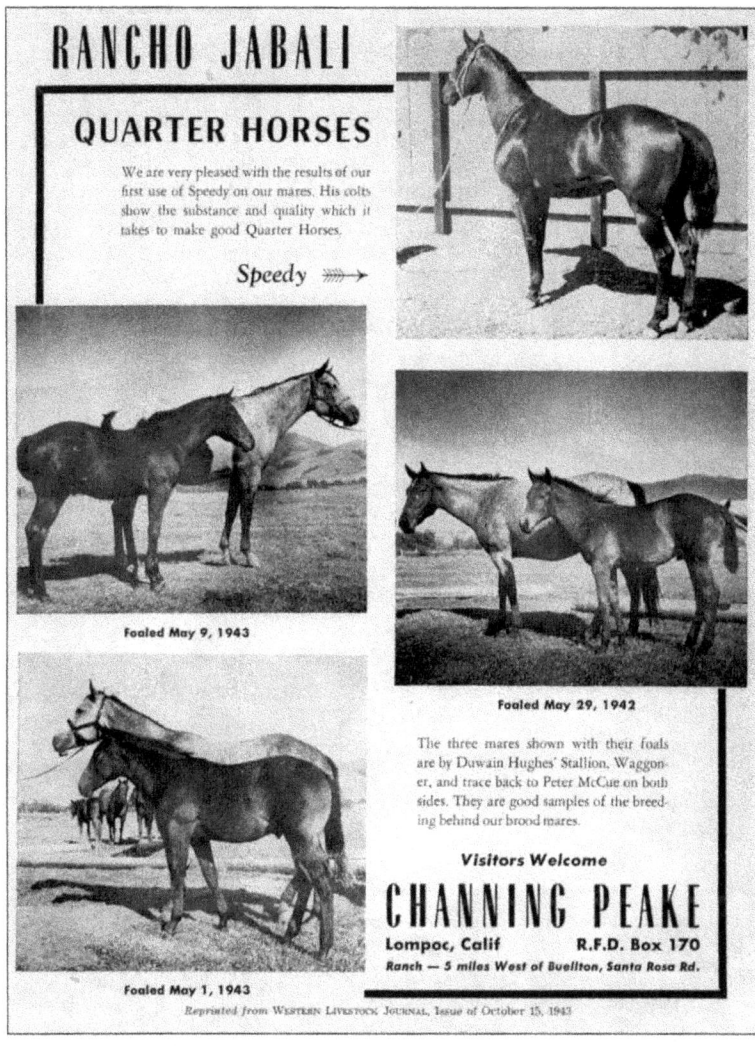

Fig. 2-15. Photo courtesy Peake files.

Advertisement in the <u>Western Livestock Journal</u>, *October 15, 1943, showing Driftwood (Speedy), mares and foals.*

ads, the Peakes listed the horse as "Speedy" rather than Driftwood. That name would come later after he was registered with the AQHA.*

An important chapter in Quarter Horse history was about to begin.

The relationship between the Peake family and Asbury Schell did not end with the sale of Driftwood. The Arizona cowboy was to purchase colts from them and find fillies which fit into the Rancho Jabali breeding program. He also helped in tracking down Driftwood's breeding so he could be registered with the American Quarter Horse Association. Throughout the years Schell's correspondence to the Peakes continued to express an interest in how Speedie was doing and the colts which he was siring. And until he hung up his ropes, the Arizona cowboy rode Driftwood horses.

Note: *For many years during Driftwood's lifetime he was called by both names, especially among the cowboys who had known him during his rodeoing days. For people not "in the know," it could be confusing.*

3

THE PEAKES AND DRIFTWOOD

"He set a record for versatility and toughness that is difficult to equal...."
—Katy Peake

Residents of Santa Barbara, California, Katy and Channing Peake entered the fledgling Quarter Horse business in 1940. They were not interested in the short, chunky "bulldog" Quarter Horses which were winning in the show ring. Their goal was to raise performance horses—athletes on whom a person could ranch, rodeo, or stock horse—with emphasis upon rodeo horses. It was a demanding area of specialization on which no one had concentrated until then.

Channing and Katy Peake were a unique couple. In addition to being dedicated breeders of performance horses, they were affluent, part of the social scene in Santa Barbara, well-acquainted with the motion picture set in Los Angeles, and deeply involved in the fine arts community of Southern California. Their children Catherine, Tuni and Michael all remembered parties, both in Santa Barbara and Lompoc, which were attended by movie stars, producers, writers and artists. Among those friends were screen writer Bordon Chase, who wrote "Red River;" actor Gregory Peck; western artists Joe DeYong and Will James; and other notables such as Colonel Tim McCoy. There was also a constant stream of horse people, cowboys, and ropers visiting the ranch. All of that company made for a varied mix of individuals who associated with the family on a regular basis.

When the Peakes decided to enter the horse business they became strong supporters of the American Quarter Horse Association, purchasing stock shortly after it was organized. They also helped form the Pacific Coast Quarter Horse Association, and Channing served as its first President. They felt deeply that "the torch" should be passed on. In 1958, they presented California State Poly-Technical College at nearby San Luis Obispo with four Driftwood daughters. Those mares were Tweedle Dum, Annie Wood, Quick A Lick, and Tweedle Dee.

Channing Peake was well-traveled and artistically inclined (he owned an original Picasso painting), a painter specializing in seascapes of gallery quality, and interested in literature. He was also a horseman, roper, a member of the prestigious Rancho Vistadores Trail Riders, and could associate with people from all walks of life on a friendly, easy basis. He was a staunch supporter of his wife's efforts to make Rancho Jabali one of the leading Quarter Horse breeding establishments in the industry.

Fig. 3-1. Photo courtesy Peake files.

Catherine, better known as "Katy," was, in Perry Cotton's words, "A born horsewoman who seemed to have an uncanny sense of what a certain mare and stallion would produce." She had ridden hunters and jumpers as a girl and knew that "Form must follow Function" in a performance horse. She never wavered from that point of view. To her, rodeo and stock horses typified the best of what a rider and mount could achieve together—although she wasn't one to push her opinions on people. That feeling was probably the reason for her dedication to breeding the horses she did. In addition, Katy was an artist (her sketches decorate pages of this book) as well as a poet and a published author of short stories and articles in such magazines as *The Quarter Horse Journal, The Western Horseman,* and *The Western Livestock Journal.* Another hobby was photography, and she took many of the horse pictures which appear in this book. Katy was the driving force behind Rancho Jabali. Beyond the ranch, she contributed effort and money to many worthy causes. Additionally, Katy Peake was a wife, mother of three active children, and a gracious, genuine hostess whom everyone liked and respected.

Katy Peake, *the lady who made Rancho Jabali work.*

Catherine Peake Webster (their daughter) wrote "No one could have been more devoted to Driftwood than my mother. She and Driftwood had a mutual admiration, each brought something vital to the other and you could see it when they were together. He was her horse!"

Webster continued "He was her horse and although she thought so from the beginning, my mother never took it for granted that everyone should think Driftwood was a great horse and sire. She wasn't one to talk him up a lot—she felt he would earn his value on his own through his offspring and this proved to be the case."

Fig. 3-2. Photo courtesy Peake files.

Channing Peake *(left) with auctioneer and team roper* **Hoke Evetts** *on the Rancho Vistadores ride.*

Fig. 3-3. Photo courtesy Peake files.

Will James, *the well-known western author and artist, was a frequent visitor at Rancho Jabali during the early 1940s.*

Fig. 3-4. Photo courtesy Peake files.

Moving cattle *at Rancho Jabali.*

"As Driftwood sired using horses my mother was very appreciative of the individuals who took those horses out and proved them, thereby giving the bloodline exposure. She never forgot those people. Throughout her life she was always aware, when given recognition for the Driftwoods, that she hadn't done it alone by any means."

The above thought was publicly expressed in the 1966 Rancho Jabali Dispersal Sale catalog when Katy Peake wrote "Over the years we have been singularly fortunate to have some of the country's finest horsemen put many horses of our breeding to public test in the rodeo arena and in the show ring. Their skills have contributed greatly to the reputation which Rancho Jabali enjoys as the home of using horses. We take this opportunity to express our particular appreciation to some few of these people with apologies for the many notable omissions."

That was followed by a list of the horsemen and women who used the Driftwoods and the capable horses which they rode.

Acquiring property at nearby Lompoc, the Peakes began to put their operation together. The ranch lay in the rolling hills of central California, just a

Fig. 3-5. Photo courtesy Peake files.

Ranch manager Jim Carr and Driftwood *pose against the California mountains. Carr was the long-time mainspring that kept Rancho Jabali functioning smoothly.*

few miles from the Pacific Ocean. That location ensured a comfortable year-round climate for growing and developing the foals as well as being close to Santa Barbara. Eventually, the operation totaled approximately 1,600 acres. Water was piped to the hill tops so the horses in each pasture had easy access. Well-planned corrals and barns were designed and built for efficient stock handling. Perry Cotton remembered it as "One of the best places to raise horses that I ever saw."

Their first horse purchase was five mares from the Greene Cattle Company (the ROs) of Patagonia, Arizona. Those mares, by El Rey RO, were strictly Quarter Horse conformation and of traditional Quarter Horse bloodlines which traced back to well-known early day sires.

The next move was to visit the Duwain E. Hughes ranch at San Angelo, Texas. Hughes was a long-time Quarter Horse breeder utilizing One Eyed Waggoner by Midnight by Peter McCue as his main stallion. The Hughes mares all carried solid Quarter Horse ancestry and functional conformation. The Peakes bought five fillies—individuals which Hughes felt were some of the best which he had ever raised.

On the bill of sale for the mares, signed by Mr. Hughes and dated September 8, 1940, the following note appears:

"These fillies are by one of my best Quarter studs, Waggoner, he being out of Midnight and Midnight out of Peter McCue.* When the fillies were loaded and pulled out, I remarked to my foreman, Sam Chumley, that was the best load of mares that ever left the Hughes Ranch and I hope that they make the same reputation in California that they have in Texas."

Over the years, the Hughes mares did, indeed, make the same reputation for Rancho Jabali, and put many ropers, cowboys and stockhorse competitors a-horseback.

With well-planned facilities, ten top mares and a defined goal, the Peakes were ready for business. There was just one problem—they didn't have the

---

*Note: *It should be remembered that many Western horse breeders use the term "by" and "out of" in the same way, meaning "sired by" or "fathered by."*

Fig. 3-6. Photo courtesy Peake files.

Three of the five RO-bred mares *with which the Peakes started their Quarter Horse program. These mares were all daughters of El Rey R.O. and out of Quarter-bred mares.*

Fig. 3-7. Photo courtesy Peake files.

Waspy 2196. *She was one of the original RO mares with which the Peakes began their breeding program. A 1940 daughter of El Rey RO and an RO mare, she produced capable performers and broodmares when bred to Driftwood. She is shown with* **Katy Peake** *after winning an early Quarter Horse show.*

Fig. 3-8. Photo courtesy Peake files.

Ray Yanez *accepting a trophy on a Driftwood during a California show of the 1950s.*

Fig. 3-9. Photo courtesy Peake files.

Driftwood (Speedy) *at Rancho Jabali.*

Fig. 3-10. Photo by Willard H. Porter.

Driftwood and Katy Peake *in a playful mood. There was no doubt that he was "her horse" and the pair seemed to have a mutual respect and love for each other.*

stallion they felt would complement their mares. For over a year they searched for that individual. He had to have a solid Quarter Horse pedigree, be of known ability as a rope horse and have already proven that he could pass on that ability. Such a horse would have more than adequate conformation, although probably not of show ring type.

Then, in the spring of 1942, Gordon Davis told them about Driftwood. How the Peakes were finally able to purchase him is recounted in Chapter 2.

In the spring of 1943, the Peakes drove to Arizona and took possession of the bay stallion. Driftwood moved to Rancho Jabali where he was to stay for the rest of his life. During the following seventeen years he made a positive impact on the using, stock and rodeo horses of America. Bred at first to the Peake mares, and then to carefully selected individuals owned by other horsemen (representing the finest performance horse bloodlines of the time), he fulfilled the goals which Chan and Katy Peake had set when they entered the Quarter Horse business.

At the time the Peakes acquired the thirteen-year-old Driftwood, he was not registered with the American Quarter Horse Association. When the stallion was foaled, there was no Quarter Horse Association (it was not formed until 1940). In 1946, because of confusion over his true pedigree, the Peakes made a trip to Texas to find out first-hand just how he was bred. They talked with old-time horsemen who remembered Driftwood and learned that he was bred by a Mr. Childress in the Silverton area. He was later owned by Sam Turner, Amos Turner, then by Ross Brinson and Buck Nichols, George Cline, and finally by Asbury Schell.

As already mentioned, the Peakes discovered that Driftwood was sired by Miller Boy and out of the Comer mare owned by Mr. Childress. Miller Boy was sired by the Hobart Horse out of the Wylie mare who was by Texas Chief by Lock's Rondo. The Comer mare was by Barlow by Lock's Rondo and out of a Kentucky Thoroughbred mare. That pedigree made Driftwood a line-bred Lock's Rondo, a horse who had set the match tracks afire around the turn of the century. They also heard that he was sired by Line Up (TB) but proved to themselves that he was by Miller Boy. Regardless of any confusion over the stallion's bloodlines, they were good and produced a horse of outstanding physical ability with the prepotency to pass it on to succeeding generations.

The visit to Texas confirmed that Driftwood (Speedy) was by Miller Boy and that is the way Driftwood's pedigree appears on the original

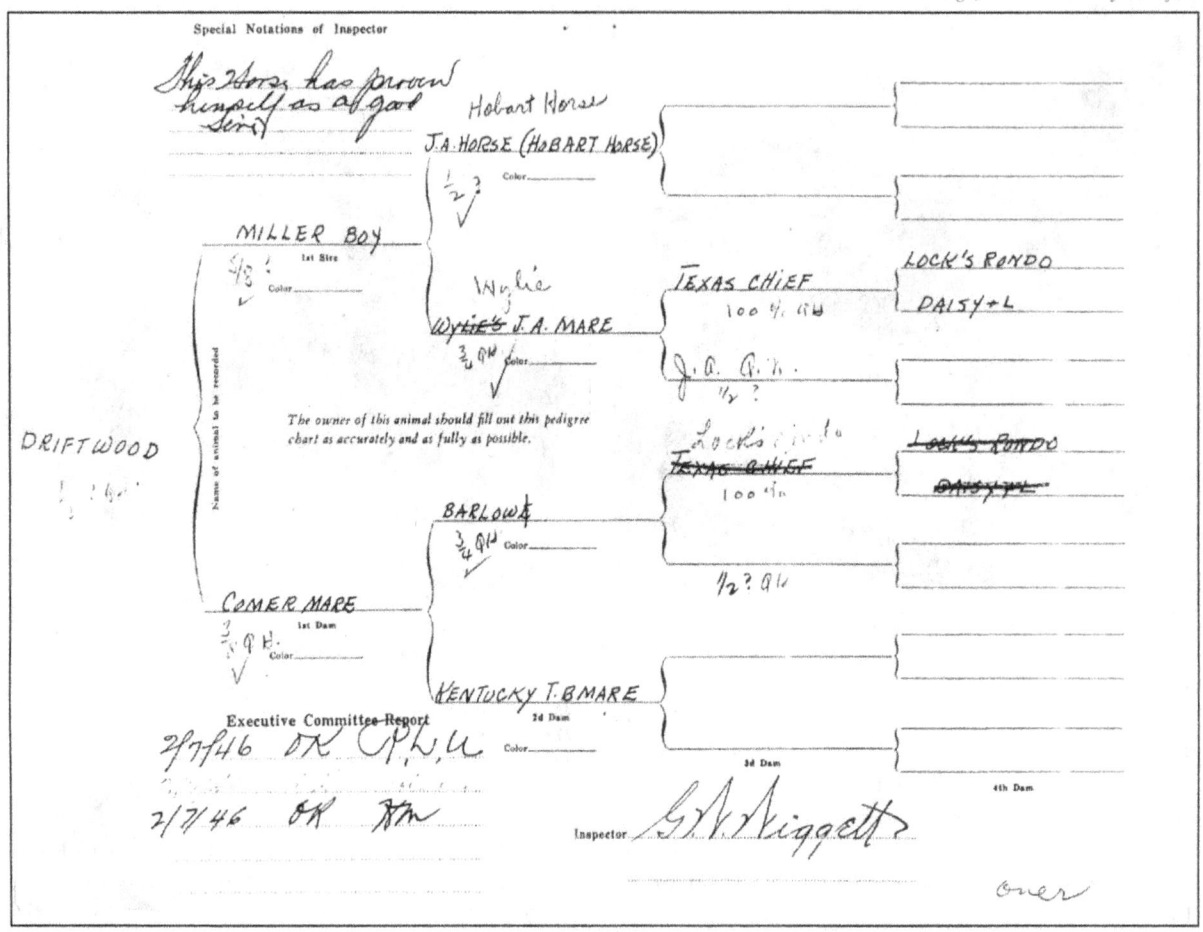

Fig. 3-11. Photo courtesy Peake files.

Registration Application to the American Quarter Horse Association.

registration application dated February 7, 1946 and signed by inspector G.W. Wiggett. R.L. Underwood, President of the AQHA and Helen Michaelis, the Secretary, also initialed the application, expressing their approval of granting a registration certificate to the stallion.

A note written on the application by the inspector expresses the opinion that "This horse has proven himself as a good sire."

During the next fifteen years that opinion was to be expressed numerous times as Driftwood's get distinguished themselves. While Rancho Jabali did not show extensively, other individuals did and ropers took his sons to the rodeo arena. Driftwood's reputation as a sire continued to grow.

According to Katy Peake, "The conformation of Speedy's get was variable. We have been careful of the mares we have bred to him and by and large they carry more Quarter Horse characteristics than he does himself. He passes on his intelligence, kindness, speed and way of going with great fidelity."

Regardless of varying conformation, Driftwood's get possessed a special look, a way of moving and a willingness to perform that distinguished them. Most horsemen of the time could "tell 'em at a glance." Even today, that same "look" makes the Driftwoods easily recognizable.

Of course, the fact that Driftwood was never bred to unbroken, high strung, or unproven mares—either the Peake's or others—is another reason why his get were characterized by excellent dispositions.

An example of that disposition took place at Rancho Jabali. Another stallion, Booger H, got loose one night and decided that there was not enough room in the stable area for two studs. Booger H

Fig. 3-12. Photo courtesy Peake files.

INCORPORATED UNDER THE LAWS OF

№ 273 TEXAS Shares 1

AMERICAN QUARTER HORSE ASSOCIATION

FORT WORTH, TEXAS

This Certifies that Catherine A. Peake is the owner of One Shares of the Capital Stock of

THE AMERICAN QUARTER HORSE ASSOCIATION

transferable only on the books of the Corporation by the holder hereof in person or by Attorney upon surrender of this Certificate properly endorsed.

In Witness Whereof, the said Corporation has caused this Certificate to be signed by its duly authorized officers and to be sealed with the Seal of the Corporation this 9th day of June A.D. 19 42

PRESIDENT SECRETARY

SHARES $10.00 EACH

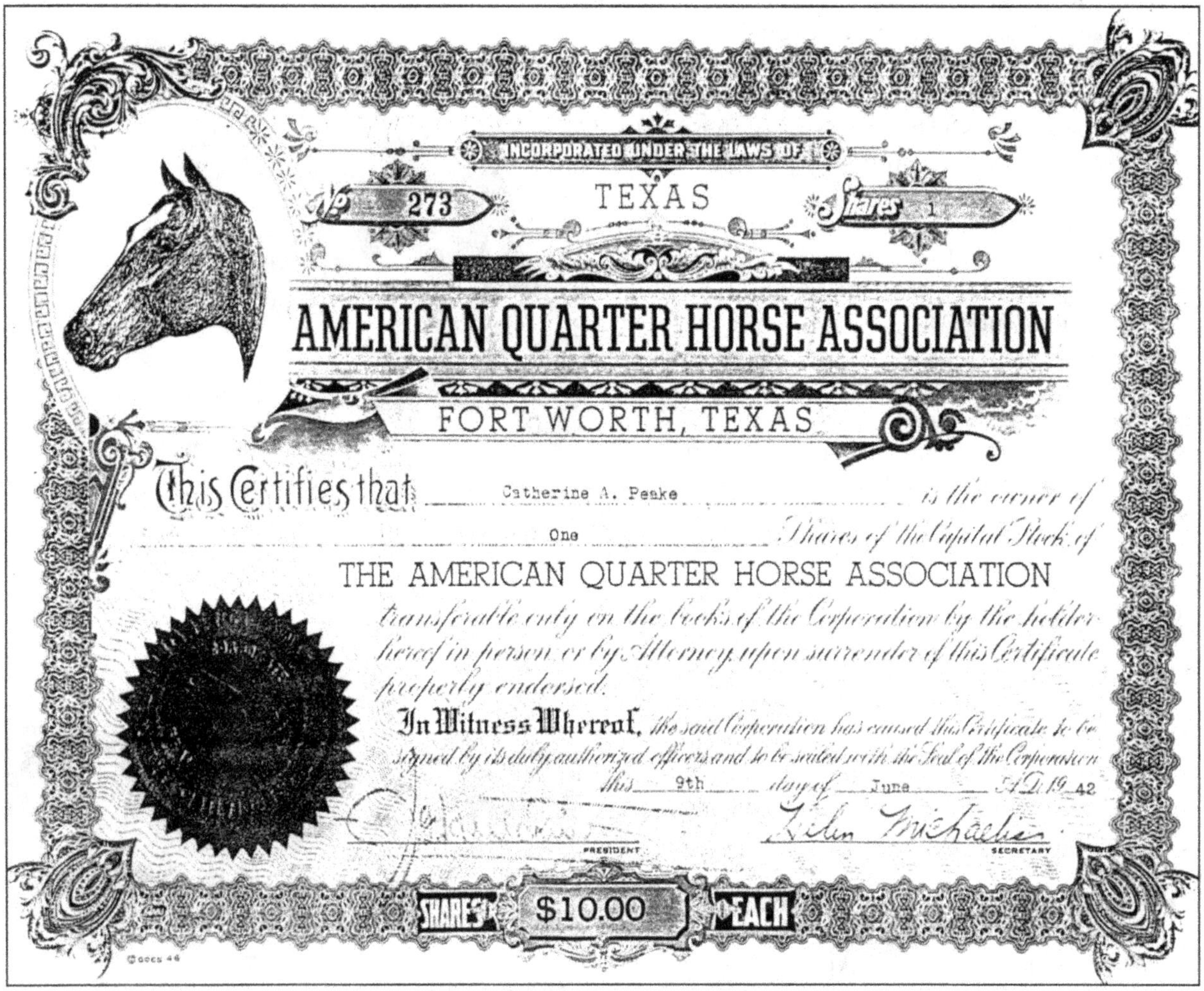

Certificate of Membership *in the American Quarter Horse Association for Katy Peake dated June 9, 1942.*

literally kicked the lock off Driftwood's stall and tore into him, biting, kicking, and squealing.

The two stallions fought in the stall and then out in the stable yard. Chan and Katy Peake weren't home, but their daughter, Catherine, was. She went out in the dark and drove Booger H off. Then, she called to Driftwood. He came to her, she slipped a halter over his head and put him away in another stall. Not many stallions would have responded in the way Driftwood did after the excitement of a fight or allowed a small girl to handle them.

Driftwood's first foals fulfilled the Peakes' hopes and found ready buyers. To identify those foals, a "P" was branded on the left gaskin. A number of Arizona horsemen who had bred to him before shipped their mares to the Pacific Coast or purchased weanlings. Some of those individuals put in standing orders for the Driftwood foal out of such-and-such a mare each year, knowing that in the future they would be well mounted.

By 1946 the demand for Driftwood get had grown to the extent that Rancho Jabali held its first

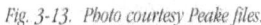

Fig. 3-13. Photo courtesy Peake files.

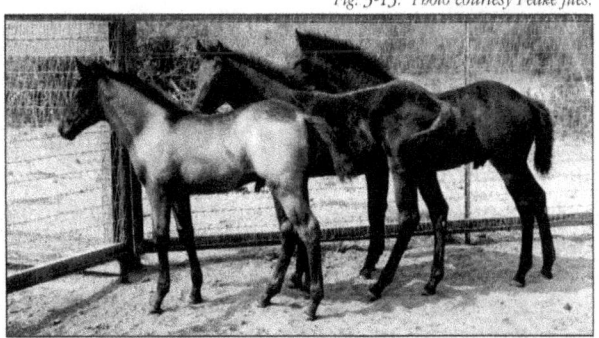

1944 foals *at Rancho Jabali. Roan foal at left is Cherokee Jake.*

production sale. That well-attended event saw such foals as Easy Keeper go through the ring. (Easy Keeper was to spend his life at the Will Gill & Sons Ranch at Madera, California and sire a host of good usin' horses.) The successes of those colts and fillies added to the interest in the Driftwoods.

Early in their program the Peakes realized what producers the King daughters were. They acquired two of them from Perry Cotton and were in the market for more. In fact, they were seriously considering a son of King to use as an outcross stallion on Driftwood daughters. In 1945 the couple went to the Jess Hankins sale at San Angelo, Texas with the thought of purchasing Poco Bueno. The brown stallion would add another "dash" of the King blood which had so impressed them. They dropped out of the bidding when E. Paul Waggoner raised his hand at $5,700. They did, however, buy several King daughters and mares in foal to King.

"That was when we changed our entire breeding program," remembered Katy. "We decided to stop off at Wilcox, Arizona and see if Old Tony could be bought. As you know, we did get him. The results of crossing him on King mares has been all and more than we hoped for."

The combination of Old Tony/King daughters bred to Driftwood resulted in the type of horse the Peakes were dedicated to breeding.

In 1947 a partnership was formed with Perry Cotton and additional mares and stallions were brought into the program. Cotton, one of the most astute horsemen of his generation, added the blood of King P-234 to the Peakes' broodmare band through such producers as Shu Cat, Queen Ann, Rocksea, Spiderette, High Tone, Red Jane C, and others. Booger H, a son of King and a full brother to Squaw H and and Hank H, was also utilized as an outcross on Driftwood daughters. Because of Perry Cotton other influential stallions were included, such as Quicksand, a grandson of Red Dog; War Chief by Joe Hancock; and Braz d'Or by San Simeon.

Perry Cotton was also to bring in the Lucky Blanton mares, combining the bloodlines of two of the greatest performance sires in the country. The resulting Driftwood/Lucky Blanton foals were popular with ranchers, ropers, and show horse folk and could "do it all under saddle." Both stallions had been successful race horses, proven rodeo mounts,

Fig. 3-14. Photo courtesy Peake files.

Old Tony *was purchased from W. D. Wear of Wilcox, Arizona in November of 1945. His daughters, bred to Driftwood, or Driftwood bred to Tony daughters, produced some outstanding performance horses.*

Fig. 3-15. Photo by Katy Peake.

King daughters *in the Rancho Jabali broodmare band. Left to right: Queen Ann, Shu Cat, Rock Sea, and Spiderette.*

Fig. 3-16. Photo courtesy Peake files.

Three of the Lucky Blanton mares *brought into the Rancho Jabali breeding program by Perry Cotton. The Driftwood/Lucky Blanton cross produced a superior usin' horse.*

Fig. 3-17. Photo courtesy Peake files.

Easy Keeper (Apple) *was purchased by Will Gill, Jr. as a yearling and stood at the Gill Ranch at Madera, California for his entire life. He sired a whole herd of top ranch and rodeo horses.*

had functional conformation and excellent minds. Today, a horse with those two names in its pedigree is sought after and demands a premium price.

Fig. 3-18. Photo courtesy Peake files.

Perry Cotton *and son "Gene" on* **Driftwood**.

Cotton remembered meeting Chan and Katy Peake in 1940 at the Los Angeles Livestock Show. The mutual interest in good horses drew them together. As members of the California Quarter Horse world, it was natural that they would become friends—and later, business associates. They would swap mares, use various stallions, hold sales, work together towards the common goal of top performance horses, and improve the quality of Quarter Horses on the Pacific Coast for fifteen years.

Perry was a native of Arkansas who had grown up in Oklahoma (a state known for good horses), made his way to California in 1933, and became a successful cattle rancher and cotton farmer. In his own words, however, "All my life, the horses have come first." He was to own, breed, and sell numerous outstanding individuals during his long career. He trained colts, roped calves and steers, showed halter and performance champions, owned and trained Thoroughbred race winners, and has been a premier breeder of fine horses for well over half a century.

The Peake-Cotton team was successful. The combination of proven bloodlines and conformation, an inborn feeling as to how those bloodlines would nick, the "horseman's eye" possessed by both Katy and Perry, and top riders on horses which would promote themselves paid off with a popular product. For over a decade, performance horsemen flocked to Rancho Jabali to purchase foals or book their mares to the stallions.

Another contributor to the popularity of the Driftwoods was Bill Gibford. A highly respected horseman, he started a number of colts under saddle for Rancho Jabali (including Cotton Cat, Speedy Peake, and Double Drift) and was well aware of the performance and producing ability of the line. As head of the Horse Department at Cal Poly, he was instrumental in adding four Rancho Jabali mares to the college breeding program in 1955. The fillies included one two-year-old and three yearlings at $375 per head for a total of $1,500. Those fillies were Cotton Cat by Driftwood and out of Katy King by King; Kitony Hancock by Roan Hancock and out of Honey Bug by Tony; Bess Shot by Big Shot and out of Peaches Carter by Rowdy; and Bobbinet Peake by Driftwood out of Bobbi Pin by Ben Hur. The stallion Braz d' Or, by San Simeon out of Lady Collidge, was purchased from Perry Cotton. It was through Gibford's efforts that the well-known Colt Training Class was included in the

Fig. 3-19. Photo courtesy Peake files.

Bill Gibford of San Luis Obispo, California. As head of the Horse Department at Cal Poly he was highly influential in adding Driftwood mares. He started a number of Driftwood colts under saddle for Rancho Jabali, including Double Drift.

Fig. 3-20. Photo by Cal Poly News Bureau.

The four Driftwood daughters presented to Cal Poly, San Luis Obispo by the Peakes. From the left: Tweedle Dum ridden by Marvin Becker; Annie Wood ridden by Glen Gimple; Quick A Lick ridden by Norbert Oliviera; and Tweedle Dee ridden by Jim Flanagan.

Cal Poly curriculum. The fillies were started under saddle by students with Gibford's direction. It was also because of the Peakes' faith in the man that they presented Cal Poly with another four young mares (Tweedle Dum, Annie Wood, Quick A Lick, and Tweedle Dee) in 1957. When bred to college-owned stallions, those mares did much to help the school's reputation as a source of good horses.

Bill Gibford not only helped to enhance Driftwood's reputation as a sire of performance horses but also influenced a generation of California horsemen through his efforts. Former students of Cal Poly who fell under his training have made their marks in the horse industry for two decades.

A survey of Rancho Jabali business statements indicates that Driftwood was standing for the fee of $100. Those statements also list the individuals who bred mares, as well as the mares which were brought to the ranch. Mare care was $1 per day. One statement to Manny Sullivan of Soledad, California, contains the note from Katy, "I don't have a record of the day the mare left. You figure the amount owed." There is a note at the bottom of the bill indicating that she received a check for $230 for the stud fee and board.

Another statement to Francis Sedgwick of Los Olivos, California lists the sale of one 5-year-old gelding, Little Boog 46258, for the sum of $650—giving an indication of the prices commanded by Rancho Jabali horses. The Rancho Vistadores rented 35 roping steers and 15 cows and calves for use during the trail ride for the princely sum of $225.

Fig. 3-21. Courtesy Peake files.

GIFT TO CAL POLY
Driftwood Daughters Will Stay Together

SAN LUIS OBISPO, Aug. 11—The rich strain of sentiment that lies deep in all true horsemen was in evidence here today: Driftwood's daughters are going to stay together.

Recently, California State Polytechnic College purchased four of the famous quarter horse sire's final crop of yearling fillies. Driftwood's owners, Mr. and Mrs. Channing Peake, who operate Rancho Jabali near Lompoc, had five of the crop left. Good offers were numerous, but the Peakes decided they would like to see Driftwood breeding perpetuated with the the great horse's final yearling fillies kept together.

As a consequence, Cal Poly's President Julian A. McPhee announced that Driftwood's five remaining daughters will join the other four as a gift to the college's home campus horse-breeding program.

The entire group, says Dean of Agriculture Vard Shepard, will be housed in the college's green and white paddocks and will be used for judging, riding and handling in the instruction program under William Gibford, animal husbandry horse specialist. They will also be used in Cal Poly's colt-handling classes and will be bred to the college's noted Bras d'Or.

Peake is an artist. His wife, who manages the ranch, has been affiliated with various motion-picture productions. They have long been interested in Cal Poly's horse program. Groups of students have frequently visited Rancho Jabali to study and practice-judge the Peakes' horses.

<u>Los Angeles Times</u> — article, August 12, 1957.

Fig. 3-22. Courtesy Peake files.

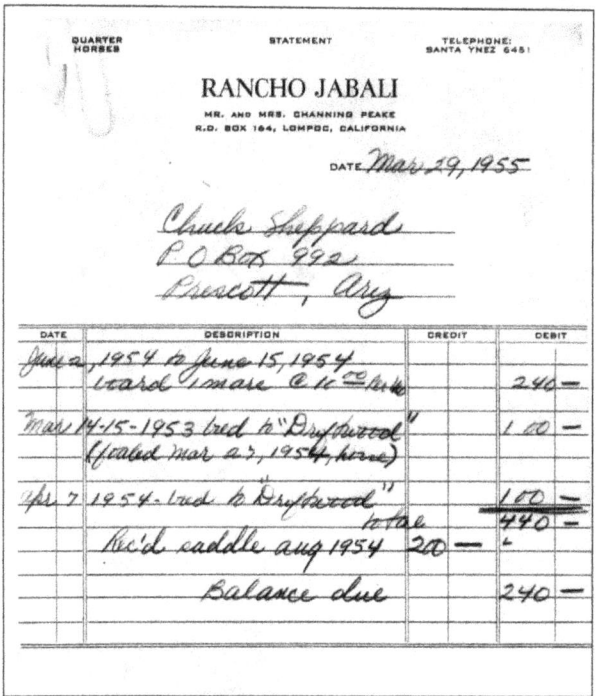

Statement *showing charges for boarding, feeding, and breeding.*

Fig. 3-23. Photo by Katy Peake.

"Old Speedy in his corral" *was the caption on this 1956 photograph taken at Rancho Jabali. At the time, "Speedy" was 24 years of age.*

On March 29, 1955, the statement shown in Fig. 3-22 went to Chuck Sheppard in Prescott, Arizona. These records, all written in Katy Peake's hand, give an intimate look at the ranch operation, how much the stallions' stud fees were, and what the horses were selling for at the time.

Channing and Katy Peake were divorced in the late 1950s and the Cotton-Peake partnership dissolved. The pace at Rancho Jabali began to slow down. The ranch still stood stallions, mares were bred and foals born, but the old excitement seemed gone. Katy was, however, too dedicated to her long-term goal to let the quality of horses drop off. Her nephew, Max Schott, had come into the operation and continued the program.

Schott was, in Perry Cotton's words, "a hand." He also brought youth and enthusiasm to Rancho Jabali, training and showing a Driftwood son, Speedy Peake, to multiple performance championships.

Then, on October 20, 1960, Driftwood died.

In a letter to her old friend Willard H. Porter the following day, Katy Peake summed up her feelings about the nearly twenty years of companionship she had enjoyed with the bay stallion.

Old Speedy is gone. He got down last night and couldn't make it up on his own.

There are no words to tell the pleasure, pride and gratitude the long association with him has given me. There will never be another like him.

According to daughter Catherine Peake Webster, "That seemed to take the heart out of my mother. She was still interested . . . but not to the same extent."

Although the event had been expected, it was still hard to absorb. Driftwood had been a member of the Peake family for seventeen years and contributed much to the success of Rancho Jabali. It was the end of an era. But Katy had a replacement waiting. She had selected Speedy Peake, a 1956 son of Driftwood out of Shu Cat by King, as the stallion who would continue the program. Under the capable handling of Max Schott, Speedy Peake developed into a versatile individual that earned ROMs in Reining, Cutting, and Calf Roping, as well as being a top contender in Working Cow Horse. As a sire, his foals proved out and carried on the Driftwood legacy.

Schott moved the operation to Klamath Falls, Oregon and put together a band of mares to complement Speedy Peake. To the time-tested performance bloodlines of Driftwood, Waggoner, El Rey RO, King, Tony, Red Dog, Joe Hancock, and Lucky Blanton, Max and Stevie Schott added Joe Reed P-3, Chicaro Bill, Midnight, Three Bars TB, Billy Anson, Braz d'Or and Oklahoma Star breeding. Mares carrying these bloodlines, when crossed with Speedy Peake, produced foals that were capable of being winners wherever they went. It looked as if the dynasty would continue for another generation.

That was not to happen. In 1966 the Rancho Jabali horses were dispersed, Speedy Peake leased, and one of the most influential breeding programs in Quarter Horse history came to an end.

Even though Driftwood and Rancho Jabali were gone, the legacy lived on. Usin' horse folk never forgot the Driftwoods and continued to breed them. Other bloodlines rose in popularity, and for most people Driftwood's name was only history—a stallion listed on a pedigree. But for the people who depended upon what a horse could do under saddle, there was no wavering in loyalty. Dedicated individuals like Roy Wales, Tex Oliver, Jake Kittle, Buck Nichols, Freeland Thorson, Will Gill, Jr., Stanley Johnson, Henry Kibbler, and Jim West kept the flame alive, and cowboys stayed mounted on the good doin' Driftwoods.

Freeland Thorson, long-time Driftwood breeder from Idaho, summed up the accomplishments of the line better than anyone with a statement to Jim Morris in 1998.

"No, you won't find a lot of AQHA points behind a Driftwood-bred horse. They don't get "pretty to the eye" until they're eight or nine years of age. But with their strong undercarriage, feet, and legs, you can count on and compete with them until they are twenty. Our Driftwoods just said to us, 'ask me if you will, I can do it.'"

During the years following his death in 1960, awareness of Driftwood's greatness as a sire intensified. His get and grandget were continuing the winning tradition that had become a hallmark of the family.

The first public tribute to the bay stallion was his inclusion in the Cow Horse Hall of Fame established by the Working Cow Horse Breeders of Yuba-Sutter

Fig. 3-24. Photo courtesy Peake files.

Speedy Peake, by *Driftwood and out of Shu Cat by King, was selected by Katy Peake to follow in his sire's hoofprints as a sire.*

Fig. 3-25. Photo courtesy Peake files.

Max Schott roping calves on Speedy Peake, by *Driftwood and out of Shu Cat. Speedy Peake was one of the greatest performance horses, winning at calf roping, working cow horse, cutting, team roping, reining, and western pleasure.*

Counties, California. The Hall of Fame came into being in 1975 and Driftwood was inducted in 1976. In addition to Driftwood, Lucky Blanton and Johnny Tivio were honored that year. A special book honoring the thirty one horses inducted was published in the early 1980s. It is a lasting memorial to horses which could perform under saddle and were able to pass on that ability.

In 1983 Driftwood was honored by his induction into the Trail of Great Cow Ponies at the Cowboy Hall of Fame in Oklahoma City, Oklahoma. He belongs there not only as a rope horse in his own

Fig. 3-26. Photo by Willard H. Porter.

Fig. 3-27. Photo courtesy Peake files.

Poker Chip Peake, by Driftwood and out of Sage Hen, with owner **Dale Smith** of Chandler, Arizona in the saddle. Not only was he a great calf roping mount, but his sensational good looks made him a standout everywhere.

A humorous advertisement listing the Cotton-Peake stallions. Horse Lover's Magazine, August-September 1952.

right, but as the sire of numerous others, including his famous son Poker Chip Peake, who lies next to the monument. The event was held during the annual convention of "The Wild Bunch," a group of old-time rodeo cowboys and cowgirls, during the National Finals Rodeo. Katy Peake was invited to the ceremony by her old friend Willard H. Porter, who was then serving as Director of the Rodeo Division of the Cowboy Hall of Fame. There is no record of whether she attended.

It was a fitting tribute that the bay stallion should be united with the memories of the men who rode him, and his get, in the rodeo arena.

Katy Peake passed away in 1998. To the end of her life, she maintained a strong faith in Driftwood and his continued progeny. She derived great pleasure in seeing the rebirth of interest, in the swell of breeders returning to the source of performance bloodlines, and from the ropers and stock horse exhibitors who took the Driftwoods to the arena and continually proved their ability.

Without Katy Peake and her belief and love for the stallion, there would have been no Driftwood story.

4

RANCHO JABALI

"One of the best places to raise horses that I ever saw."

—Perry Cotton

As with all horse operations, life at Rancho Jabali was characterized with ups and downs. Daily care of the horses, the bustle of the breeding season, keeping fences fixed, putting up hay, weaning and halter breaking foals, and starting two-year-olds under saddle kept everyone busy. There was also a constant stream of visitors to look at the stallions, mares, and foals, and purchase young stock.

Rancho Jabali was home to the Peake family and there was the bustle that three young children could generate. While the horses were the center of interest, there was also "a lot of living" there. Katy supervised her children, ran the home, served as hostess, wrote and sketched in her spare time. She was general manager and handled all the necessary correspondence for the ranch. Chan painted regularly (a newspaper clipping calls attention to a one-man show of his paintings scheduled at a Santa Barbara art gallery), was involved in the Southern California fine arts scene, helped run the operation, roped, and participated in outside horse industry affairs. There was also an active social life with a varied circle of friends and acquaintances from Santa Barbara and Los Angeles.

Fig. 4-1. Photo courtesy Peake files.

Buckwood, *by Driftwood out of Hancock Belle, as a 2-year-old with Rancho Jabali foreman* **Jim Carr** *in the saddle. Buckwood went on to become a show horse, rope horse, and a top producing broodmare.*

Important to the success of the operation was foreman Jim Carr. He was a life-long horseman who took a deep personal interest in the breeding program and the way in which the foals "broke out." If Katy Peake was the driving force behind Rancho Jabali, Carr was the wheel which kept it running. He was involved in every aspect of the program and intensely interested in the progress of the horses after they left the ranch. Upon his retirement, about 1950, Chan and Katy Peake presented him with a Driftwood son out of Hancock Belle. This colt, named Maestro, was never registered with the AQHA. Maestro eventually ended up in Arizona, carrying the saddle of Eddie Schell, Asbury's son.

A major source of conversation was the development of various foals bred at Rancho Jabali. Who owned them, how they broke out, and what their accomplishments were under saddle were of constant interest to everyone. Since the majority of the horses went to "good hands," the Peakes and staff usually had something to talk about.

A Driftwood which created a lot of laughs was "Monster." This colt was, according to Catherine Peake Webster, a "catch colt" by a son of Driftwood and an aged shetland pony mare.

"It was our custom to turn a gentle older horse, in this case a Shetland pony mare, in with the weanlings to further their social education. Back in those days we weaned very late and one of the colts proved to be more mature than expected.

"As ponies tend to be easy keepers we were somewhat concerned that the little mare was getting awfully fat but didn't think much about it. Were we surprised when this aged maiden pony mare foaled!

"From the beginning Monster, named by my mother, had a normal (horse-size) head which gave the rest of his conformation a grotesque aspect. Knocking Monster in the head was never even considered, so the little horse remained an ongoing source of humor on the ranch.

"When the time came, Monster was broke and seemed to have a lot of Driftwood spirit and heart. He did his best to keep up with the horses and I always enjoyed riding him."

There were also accidents, sickness and near tragedies to contend with.

Catherine Peake Webster recounted another story. She and her sister and brother were subject to regular vaccinations for tetanus (lockjaw), as were most children. Sometime during those years (probably during the late 1940s) Driftwood contracted tetanus. The reason was never determined, but according to Channing Peake, "It was because he was the best horse on the ranch."

Fig. 4-2. Photo courtesy Peake files.

Catherine Peake on Driftwood. *He was in his teens at this time, but still retained the solid conformation and alert look of his youth.*

Fig. 4-3. Photo courtesy Peake files.

A young Catherine Peake Webster on "Monster," *a grandson of Driftwood out of a shetland pony mare.*

Fig. 4-4. Photo courtesy Peake files.

Actor Gregory Peck, *an unidentified companion, and trainer Ray Yanez line up for a photo at Rancho Jabali.*

Fig. 4-5. Photo courtesy Peake files.

Channing Peake *(left) presents a trophy saddle to* **Doad Hex** *at a Rancho Jabali roping.*

Fig. 4-6. Photo courtesy Peake files.

Ralph Camarillo *(center) and his sons* **Leo** *and* **Jerold** *in the Rancho Jabali arena, 1953. Both boys were to become World Champion PRCA Team Ropers.*

Fig. 4-7. Photo courtesy Peake files.

Olin Simms, *former World Champion Team Roper, shares a laugh with* **Raymond Cornelius** *at a roping. Both men were close friends of Channing and Katy Peake and visited the ranch frequently. Simms even worked there for a while.*

Katy Peake was determined to save her favorite horse. Driftwood was moved to a dark stall and outfitted with a sling suspended from overhead beams. The important thing was to keep the stallion on his feet and somehow get fluids and nutrients into him.

Mrs. Peake took small amounts of oat hay and steeped it in boiled water which was heated on the kitchen stove. This "hay tea" was to save Driftwood's life.

A small table holding a flashlight, notebook, funnel, tubing, watch, and a large bottle warmer was set up in one corner of the stall. A folding chair was next to the table and a low-watt light bulb was screwed into a nearby socket.

At regular intervals the warm broth was poured into the funnel and then to Driftwood's mouth via the tube. The intelligent stallion caught on immediately, swallowing without difficulty. The routine was to continue 'round the clock.

Sleeping bags were moved into the stall and the Peake family took up residence for the duration of the emergency. The situation went through cliffhanger to happy ending. Driftwood finally recovered, due to Katy Peake's determination not to lose him.

Catherine remembered her father walking down to the barn, surveying the arrangements and commenting, "It looks as if everyone in the family has moved down here. I'm going to sleep in my bed at the house."

The roping arena and herd of steers on the ranch was more than a place for training horses or practicing the sport. Weekend jackpots drew many well-known ropers and were frequently the hub of activity. Ralph Camarillo and his sons Jerold and Leo, the Yanez clan, Hoke Evetts, Raymond Cornelius, Gordon Davis, Harlan Brown, Doad Hex, Olin Simms and others were all regular competitors. The Peakes even provided steers and chute help for a roping that was a part of the Rancho Vistadores trail ride one year.

A few of those ropers became ranch employees. Ralph Camarillo worked there and his sons, Leo and Jerold, began their roping careers in the Rancho Jabali arena. Both boys were to later earn

PRCA World Championship titles as professional team ropers. Olin Simms, another World Champion roper, was ranch manager for a time. Ray Yanez, a respected Pacific Coast horseman, trained and showed Driftwoods for Rancho Jabali. Walt Mason, a capable trainer, also added his touch to the Driftwood horses and showed such capable campaigners as Wooden Nugget.

Because of their acquaintance with the motion picture community in Los Angeles, the Peakes made the ranch available as the setting of at least one movie (title unknown). Slim Pickens, who gained fame as a character actor, was just beginning his career and had a part in the production. Hoke Evetts, a good friend, roper and auctioneer from Hanford, also acted in it.

"The director had me doubling a woman and there I was, in a dress, padding in the appropriate place and wig," laughed Evetts. "I even had to shave off my mustache."

The relationship with movie people also resulted in the feature film "The Bay Lady" and Driftwood's exposure to millions of viewers. Larry Landsburg (he owned at least one Driftwood) was producing animal films in connection with Walt Disney during the late 1950s. He conceived the idea of a "girl and her Quarter Horse" film and asked his friend, Katy Peake, for help.

The story briefly traced the development of the Quarter Horse from Colonial times to the 1950s, calling attention to both Driftwood as a sire of performance horses and the King Ranch strain. Most probably, no Quarter Horse stallion had his

Fig. 4-9. Photo courtesy Peake files.

A scene from "The Bay Lady" *in which Elaina Vasquez (Sammy Fancher) rescues the mare (Henny Penny Peake) and her foal from high mountain snows.*

Fig. 4-10. Photo courtesy Peake files.

Larry Landsburgh *taking a break during the filming of "The Bay Lady." Note the rustic stool.*

Fig. 4-8. Photo courtesy Peake files.

Rex Allen and Sammy Fancher, *stars of "The Bay Lady," share a laugh.*

Fig. 4-11. Photo courtesy Peake files.

Movie actor Colonel Tim McCoy and a young admirer a-horseback at Rancho Jabali. McCoy was a frequent visitor to the ranch.

Fig. 4-12. Photo courtesy Peake files.

Channing Peake building a loop at one of the Rancho Jabali ropings.

name in front of as many people as Driftwood did when "The Bay Lady" aired on national television.

The story line for "The Bay Lady" was a young woman's desire to maintain a strain of horses which her family had raised for generations, and her struggle to find and keep a mare of that breeding. Her dream was to combine the bloodlines of Driftwood and the King Ranch horses. She not only found the mare she wanted but was able to eventually breed her to the well-known Hired Hand's Cardinal. The good Driftwood daughter Henny Penny Peake was chosen for the title role as "The Bay Lady." Under Jimmy Williams, she had already distinguished herself as one of the premier hackamore and bridle horses on the Pacific Coast. Her classic looks and adaptable attitude made her a natural for the part. Sammy Fancher Brackenberry, a well-known horsewoman, World Champion Barrel Racer, actress, and stunt woman, was given the human lead as Elaina Vasquez. Ranch-raised Rex Allen, western singer and actor, narrated and had a part in the production as the man who helped Elaina find her mare. Katy Peake served as technical advisor to make sure that the events depicted were accurate from a horseman's viewpoint.

Produced in 1957, "The Bay Lady" was shot on the California coast and in the Sierra Nevada mountains, the Gitzwaller ranch and the Sonoita fairgrounds in Arizona, and the Waggoner and King Ranches in Texas. In addition to Henny Penny Peake, such well-known Quarter Horses as Poco Bueno and Hired Hand's Cardinal were shown. Jimmy Williams, E. Paul Waggoner and Robert Kleburg also appeared in cameo roles. Another rancher who had a brief part was Abe Graham. For him, it was a nostalgic return to his youth, since he had owned and roped off of Cowboy Schell (by Driftwood) prior to selling him to Asbury Schell fifteen years before. The cattle working scenes were photographed in Arizona and both Waggoner and King Ranches in Texas. Of special interest to horsemen was Jimmy Williams putting the "Bay Lady" through her paces with just a leather string in her mouth, the Quarter Horse mare Jodie Earl cutting cattle without a rider, and scenes of both the Waggoner and King Ranch remudas and broodmares.

"The Bay Lady" appeared on national television during "The Walt Disney Show" and received wide acclaim. One viewer commented, "I've never seen so many good Quarter Horses in such a short time. And, well, Sammy and Henny Penny Peake just fit."

Sammy Fancher Brackenberry recalled traveling around the county with the film crew, shooting at various locations. "When we originally discussed the story the producer needed a name for my character. I suggested Elaina Vasquez, my grandmother's name. She was born in Monterey, California, into an old ranching family before the turn of the century. It fitted right in with the story."

During an interview with Jim Morris, Brackenberry laughed about the problems the cameraman had with her makeup during the snow scenes. "I'm dark-complected, and with the light reflecting off the snow, I looked almost black on film. The makeup artist had to really lighten me up for the reshoot." Sammy also joked about one scene in which she had to rope a calf. "It took thirty takes. I don't know why they shot that many because I never missed a loop."

As mentioned earlier, the relationship with Asbury Schell did not cease after Driftwood was purchased. Occasional correspondence from the Arizona cowboy showed that he still maintained an interest in what the Peakes, and Driftwood, were accomplishing.

Fig. 4-13. Photo courtesy Peake files.

Visitor Charlie Mickle, who was to become a successful Quarter Horse breeder in his own right, gets a seat on **Driftwood** in 1954. Mickle owned and roped on Bill Adair by Driftwood and out of Mayflower.

June 11, 1943
Mr. Channing Peake,
Dear Friend:

I guess you will be surprised to get this letter. I am going to try my hand at writing one time anyway. How is Speedie and the rest of the family? Fine, I hope. I suppose all the little Speedies have arrived by this time. Please write and tell me about them. I left Tempe about two weeks after you went home. We moved to Coolidge. My brother and I have been running about 5,000 head of cattle for Tourea Packing Co. We have really been busy. Edward is the only help we have had. Believe me these Mexico steers are sure hard on these broken down fences. It is sure windy over here.

I haven't been able to find you any Clabber fillies up to date that I think would suit you. I have one two year old, she is sure pretty, but I haven't decided yet if I want to keep her or not.

Eddie is disappointed that his Speedie colt was a filly. Please give my regards to your wife, Dudley, Anna and Jim.

Best Regards to all,
Asbury Schell

~

July 18, 1943
Mr. Channing Peake,
Dear Friend:

Rec'd your letter & picture. I sure thank you a lot for the picture. I sure thought it was a good one. I would like to see the colts, but I can't, so I will just pick Smokey's it looks alright to me. I can send you the money now, or pay you when I come over after it, just which ever you wish. We sure thought that was a good picture of Speedie. Dorothy tacked it up on the wall. I would have liked to have seen you after that calf.

My brother Joe has a colt coming about the 1st of August from Speedy and a Clabber filly. One of the mares I bred for Chester Cooper from up at Roosevelt had twin horse colts from Speedy, but they came about six weeks too soon and they both died.

Have you sold any of the colts that were born last year? I

Fig. 4-14. Photo by Wolf Lauter; courtesy Peake files.

Ray Yanez *collects another win on a Driftwood.*

Fig. 4-15. Photo courtesy Peake files.

Western artist Joe DeYong, *a friend of the Peakes, mounted on* **Bootlegger**—*the last rope horse of Will Rogers. The pair is probably lining up for the Old Spanish Days Fiesta Parade at Santa Barbara.*

don't believe you will have much trouble selling Speedy's colts.

Was Sage Hen's colt a horse colt? It ought to be a good one. I will try and get over there when the colt is old enough to wean and see you folks. We want to show you people around over here when this war is over. There is no news over here so will say goodby.

Best Regards
Asbury Schell

A letter the summer after he had sold Driftwood (Speedie) verified that he had not forgotten what a calf roping mount the bay stallion was.

August 26, 1943
Mr. Channing Peake,
Dear Friend:

I am writing to ask a very big favor from you. I am about half way in the notion to go to Madison Square Garden this fall, but don't know if I can get away or not. If I can go I would like to borrow my old Speedie horse.*

If you feel like you shouldn't loan him just say no, for I know how it is. I heard about a Clabber filly that I will look at as soon as I can get to it, if you still want one. How are the colts getting along at this time?

Don't be afraid to say no about Speedy. I don't know for sure if I can go anyway. If I do go we will have to leave around the 25th of September, and I ought to work him a couple of weeks. Please give my regards to everybody.

Respectfully,
Asbury Schell

November 8, 1943
Dear Friends:

We rec'd your most welcome letter a few days ago. Was surely glad to hear from you. We liked the picture of the colts very much. Surely do appreciate the trouble you are going to for me to pick up a colt.

*Note: *There is no record if the Peakes told Asbury Schell that he could use Speedie (Driftwood) at New York that fall or if Schell even was able to compete. It was interesting to know that, of all the horses the roper could have gotten a seat on, he still felt that Driftwood was the best.*

Fig. 4-16. Photo courtesy Peake files.

Trainer Elwin Hall *prowls the pastures with a pair of Driftwoods.*

Fig. 4-17. Photo courtesy Peake files.

The cutting arena *at Rancho Jabali. The ranch was located in the rolling, oak-studded hills of California, just a few miles from the Pacific Coast. The rolling terrain helped give the foals their good feet and legs.*

Just looking at the pictures I like the looks of Sage Hen's colt the best. I doubt if we can take the time to go to the L.A. rodeo, but would surely like to. Would also love to visit you folks, but am afraid we will have to make a hurried trip. I will try and get over there about the 1st of December and if possible would like to visit you folks a day or two.

The fellow who has that Clabber filly wants $300 for her. She is a nice filly coming two years old. My opinion of her and yours might differ so I will not urge you to buy her without seeing her.

Yours truly
Asbury Schell

~

November 13, 1943
Dear Asbury,

We were glad to hear from you and know that you are planning to come out here after your colt. We are pretty sure that you are going to like Sage Hen's colt the best. He is without any doubt the most like Speedy. In fact, except for a little white star on his forehead he is a dead ringer for his pappy, color and all.*

If we get down to the Stock Yard Show it will be on the 27 and 28th and we will surely be home by the first. Be sure to let us know as soon as you can just when you will be here.

We have decided to take a chance on the Clabber filly and I am enclosing a check made out to you so you can pay for her. I presume you will bring your trailer with you so you can take your colt back with you. If it isn't too much extra trouble could you pick up the filly and bring her along with you? Otherwise there won't be much sense in our buying her as we couldn't possibly get the gas for such a long trip. Please don't feel responsible about our picking out a filly on your judgement. If we are disappointed in her it will be just our own hard luck.

Do you suppose that the Nichols would have any information on Speedy's breeding? We haven't had any luck trying to track it down and we are very anxious to get him registered.

Our best to all of you.
Channing and Katy Peake

~

December 9 1943
Mr. & Mrs. Channing Peake,
Dear Friends:

Will write to let you know that we arrived here safe and sound. The colt never hardly moved all the way over here. It showered on us a little after we hit Arizona.

I will send you the address of Ross Brinson, the fellow that brought Speedie to Arizona. It is P.F.C. Ross Brinson, 1852nd Hdq. Det., Fort Bliss, Texas. Tell him I mentioned

---

* **Note:** *The Driftwood/Sage Hen colt which Asbury was to purchase was Drifty P-6668. Under Eddie Schell, he became a top team roping mount and carried his rider to a World Championship in the event.*

Fig. 4-18 Photo courtesy Peake files.

Speedy Bar by Driftwood, with **Tuni Peake** up. In two years he won more than 100 ribbons in hunter, jumper, equitation, and pleasure classes. Over half were blue.

you folks writing him. He is a very nice accommodating kind of fellow, and I am sure he will help you all he can.

Best regards to all
Asbury Schell

P.S. give Jim my regards

May 4, 1946
Mr. Channing Peake,
Friend Chann:

Guess you will be surprised to get this letter from me after so long a time. How is the ranch and all your horses?

I am writing to see if you have Speedie registered yet? And if so I wish you would send me all the information you can as I would like very much to get Drifty registered.

He is getting to be quite a horse. I didn't show him at Tucson as he had a bad case of distemper. I guess you heard I bought the Cowboy horse from Abe Graham. I was about to get some information for you on that Clabber filly but the guy that brought her mother out here committed suicide. Drifty has a new son only 9 days old. It is sure a cute colt.

I am going to rodeo for a couple of months this summer starting about the 4th of July. Eddie has been doing pretty good in team tying here lately. I would appreciate it a lot if you would help me get Drifty reg.

Fig. 4-19. Photo courtesy Peake files.

Two of the Rancho Jabali mares and a foal line out across a pasture.

Best regards to all
Asbury Schell

July 20, 1946
Mr. Channing Peake
Dear Friend:

No doubt you will drop dead with surprise when you get this letter. I've been putting off for a long time. How is everybody?

How is Speedie? Has he lots of sons and daughters this year? I believe the colt I bought from you is the best looking colt I have ever looked at. He weighs about 1025 lbs. He won't be as tall as his dad but he will be heavier.

I have heard you have Speedie registered in the Quarter Horse class. If so I wish you would send me all the dope, as I would like to have Drifty registered.

Have you sold all of his colts this year? If not, I would like to get another one. I wouldn't mind having a full brother to Drifty. What are you getting for them now?

Eddie's colt and the other colt are not near as good as Drifty. I want to take him to Tucson to the horse show next year.

Well we will be seeing you when this war is over. Will appreciate it if you will send me the dope.

Respectfully yours,
Asbury Schell

Fig. 6-5. Photo courtesy Peake files.

Shu Cat P-5619, black, foaled 1943 by King P-234 out of a Hankins Quarter mare. Her Driftwood foals included:
>**Driftwood II,**
>**Peakewood,**
>**Speedy Scat,**
>**Speedy Peake,**
>**Kittywood**
>**Runaway Peake.**

Fig. 6-6. Photo courtesy Peake files.

High Tone P-2189, sorrel, foaled in 1942 by King P-234 out of Uncle's Pet 2505. Her only Driftwood foal was:
>**Fleet Wood.**

Fig. 6-7. Photo courtesy Peake files.

Spiderette P-13359, black, foaled in 1942 by King P-234 out of Spider H. P-13102 by Darity TB. Her Driftwood foals were:
>**Swiftwood,**
>**Jernigan Peake**
>**Wood Mite.**

Queen Anne 2781, chestnut, foaled in 1942 by King P-234 out of a Holland mare by a son of Harmon Baker. By Driftwood she foaled:
- Miss Linwood,
- Brown Beulah
- Annie Wood.

Fig. 6-8. Photo courtesy Peake files.

Rock Sea P-3232, by King and out of Silly by Show Boy. She was bred to Tony, producing:
- Coaster.

Fig. 6-9. Photo courtesy Peake files.

Nugget Hug P-8456, sorrel, 1944, by Bear Hug 2868 out of Goldienug 8451 and bred by Sid Vail of Douglas, Arizona. By Driftwood she produced:
- Hug Me Tight,
- Quick A Lick
- Wooden Nugget.

Fig. 6-10. Photo courtesy Peake files.

Fig. 6-11. Photo courtesy Peake files.

Hancock Belle P-5593, dun, foaled in 1944 by Buck by Chicaro out of Miss Hancock by Joe Hancock. Her Driftwood foals included:
  Buckwood,
  Maestro,
  Chilena,
  Driftwood Ike
  Speedywood.

Fig. 6-12. Photo courtesy Peake files.

Dusky Ruth 34067, sorrel, foaled in 1949 by Lucky Blanton out of Bajita Fifi by Hard Tack. When bred to Driftwood she foaled:
  Dusky Peake,
  Speedy II,
  Speedy Darnell
  Woodmist.

Fig. 6-13. Photo courtesy Peake files.

Sweet Hallie 2689, bred by Jack Casement by Red Dog out of Buckeye. She foaled:
  Chipperwood 55344 by Driftwood.
  Chipperwood was first owned by Bordon Chase, then by Roy Patton, and finally by John Hoyt, where he sired a number of Working Cow Horses that set the show ring afire.

Speedy Bar by Driftwood, with **Tuni Peake** up. In two years he won more than 100 ribbons in hunter, jumper, equitation, and pleasure classes. Over half were blue.

Two of the Rancho Jabali mares and a foal line out across a pasture.

you folks writing him. He is a very nice accommodating kind of fellow, and I am sure he will help you all he can.

Best regards to all
Asbury Schell

P.S. give Jim my regards

May 4, 1946
Mr. Channing Peake,
Friend Chann:

Guess you will be surprised to get this letter from me after so long a time. How is the ranch and all your horses?

I am writing to see if you have Speedie registered yet? And if so I wish you would send me all the information you can as I would like very much to get Drifty registered.

He is getting to be quite a horse. I didn't show him at Tucson as he had a bad case of distemper. I guess you heard I bought the Cowboy horse from Abe Graham. I was about to get some information for you on that Clabber filly but the guy that brought her mother out here committed suicide. Drifty has a new son only 9 days old. It is sure a cute colt.

I am going to rodeo for a couple of months this summer starting about the 4th of July. Eddie has been doing pretty good in team tying here lately. I would appreciate it a lot if you would help me get Drifty reg.

Best regards to all
Asbury Schell

July 20, 1946
Mr. Channing Peake
Dear Friend:

No doubt you will drop dead with surprise when you get this letter. I've been putting off for a long time. How is everybody?

How is Speedie? Has he lots of sons and daughters this year? I believe the colt I bought from you is the best looking colt I have ever looked at. He weighs about 1025 lbs. He won't be as tall as his dad but he will be heavier.

I have heard you have Speedie registered in the Quarter Horse class. If so I wish you would send me all the dope, as I would like to have Drifty registered.

Have you sold all of his colts this year? If not, I would like to get another one. I wouldn't mind having a full brother to Drifty. What are you getting for them now?

Eddie's colt and the other colt are not near as good as Drifty. I want to take him to Tucson to the horse show next year.

Well we will be seeing you when this war is over. Will appreciate it if you will send me the dope.

Respectfully yours,
Asbury Schell

October 8, 1946

Mr. & Mrs. Channing Peake
Dear Friends:

How is everybody by this time? Fine I hope. Hadn't heard from you so thought I would write and see if you had sent me the papers on Drifty. Thought they might have gotten lost.

I was over at the Hayward Rodeo. I intended to go by and see you folks but got in a big rush to get home, so came straight back.

How is Speedie? Heard you took the mares and pony over in the valley. We haven't any cattle here now. Don't know what I will do. May buy a place around Chandler.

If you haven't sent those papers would you do so as soon as possible, as I would like to get him registered.

Best Regards to all,
Asbury Schell

~

With an eye towards promoting Driftwood, Channing Peake wrote to Schell inviting him to the Pomona, California Quarter Horse Show. Both Drifty and Cowboy would be excellent examples of what the Driftwoods could do under saddle.

February 14, 1949

Dear Asbury:

Our spring Quarter Horse show is going to be held at Pomona on May 21st and 22nd. The gelding classes are getting to be bigger and pretty important. We would sure like to win it with a Speedy colt and wonder if there is a chance of your bringing Cowboy and Drifty out here for the show. The entries close on April 15th. Let me know as soon as you can so that we can make arrangements.

We were surely disappointed not to see you this last rodeo season and really want you to come by to see a few of the good colts old Speedy is getting for us. We bought back the 1944 blue horse colt out of Pigeon*. It cost us $2,500 to get him back but we sure do like him and think that he is our best bet so far to replace the old horse.

Speedy is still in wonderful shape. His hocks never got quite right after he hurt them down in Tucson and we do not ride him a great deal any more. We have sixteen

---

* **Note:** *This foal was Cherokee Jake, who eventually went to Asbury Schell and finally to John Clem.*

outside mares on the ranch now to breed to him and what with the mares to come later on in the year plus our own he will have a really full season.

Our very best to you and your family and hoping to hear from you soon.

Sincerely,
Chan & Katy

~

As always, Asbury Schell took his time answering, but answer he did. It is interesting to note that he wanted to combine the horse show with the roping events during the rodeo held at the same time.

April 7, 1949

Mr. & Mrs. Channing Peake
Dear Friends -

How are things with you folks? fine I hope.

Fig. 4-20. Photo courtesy Peake files.

Jake Kittle, *Quarter Horse breeder and family friend, takes a walk with* **Tuni Peake** *at Rancho Jabali.*

Eddie and I will be at Pomona if nothing unexpected happens. I can't think of a thing now to keep us from being there.

Do you know when the entrys close in the roping? If you do, please let me know when, and the person to send them to.

Cowboy and Drifty are fine, but Cowboy is getting a little stiff. Too much racing I guess.

Best Regards
Asbury Schell

~

No records have been found verifying how either Cowboy or Drifty did in the gelding class or if the Schells won any money in the calf roping and team roping at Pomona.

By the time of the last letter, Driftwood's foals were beginning to make a name for themselves in the stock horse world as well as with ropers. A steady stream of buyers were purchasing foals by the bay stallion, or out of his daughters, and winning with them.

While not a regular exhibitor, Rancho Jabali did show to a limited degree, exposing the Driftwoods to the show ring and concentrating on events which would demonstrate the athletic ability and adaptable mind. Wooden Nugget, exhibited by Walt Mason, did exceptionally well in the working cow horse and stock horse events. In fifteen outs over two years, "Woody" placed first seven times, winning three trophy saddles; second four times; third once; and fourth three times. The versatile Speedy Peake, with Max Schott in the saddle, could do it all—working cow horse, roping, reining, cutting, and western pleasure—in addition to being a pro-caliber calf roping mount. In 1965, Speedy Peake won the $1,000 Championship Working Horse Stake at Santa Barbara, winning first in three of the five events. (The performances of both of these horses will be covered in their biographies later in this book.) Ray Yanez took several of the Driftwoods to the show ring, winning regularly in the cattle and reining classes. Speedy Bar, ridden by Tuni Peake, earned over 100 ribbons in hunter, jumper, equitation, and pleasure classes in just two years of showing. Over half of those ribbons were blue.

There was no doubt that Katy Peake's goal to breed and raise horses that could do something under saddle was proven out.

5

OTHER STALLIONS AT RANCHO JABALI

"We had been searching for several years but had not found the stallion to head our program. Then Driftwood appeared. After his first foals, we realized that we needed other stallions to breed to his daughters or to produce mares to breed to him."

—Katy Peake

Bomber

Tony

El Greco

Quicksand

Sun Raider

Booger H.

War Chief

Roan Hancock

Braz d'Or

While Driftwood was "The Horse" whom Chan and Katy Peake chose to head their program, they did use other stallions. They were crossed back on Driftwood daughters or he on theirs to continue the "performance plus" goal of Rancho Jabali.

The first stallions Rancho Jabali used on the original mares were of Jack Casement breeding. Bomber, Quicksand and Sunraider all traced back to Ballymoony and Red Dog—truly "foundation breeding" since the Casements had been breeding Quarter Horses for many years. Additional stallions selected to cross with the Driftwood daughters or produce broodmares were all of proven performance bloodlines and Quarter Horse conformation.

Early in the program the Peakes realized what producers their King daughters were and acquired several stallions sired by him, by purchase or by outside use. They even considered adding Poco Bueno to the program but did not purchase him. Tony, a line-bred Traveler, was acquired instead, utilized with great success, and changed the direction of the breeding program. Later, Joe Hancock blood was infused through War Chief (eventually given to Tex Oliver) and Roan Hancock, whose influence was very limited due to his death shortly after arriving in California. Braz d'Or, owned by Perry Cotton, brought into the program another dash of Traveler.

While Driftwood was the major impact upon the Rancho Jabali program, the Peakes were not blind to the fact that the influence of Traveler, King and Joe Hancock would enhance the performance ability of their horses.

Prior to the purchase of Driftwood the Peakes bred the RO and Hughes mares to **Bomber 916** *(photo not available)*, a Casement-bred stallion by Frosty 3610 by Baileymooney and out of Sanchia by Little Joe Springer.

Fig. 5-1. Photo courtesy Peake files.

Tony P-776 was bred by Jim Kenny of Bonita, Arizona and owned by W. D. Wear for most of his life. He was a sorrel by Guinea Pig by Possum (Little King) by Traveler and out of a mare by Bulger by Traveler, making him a line-bred Traveler. Foaled in 1925, he spent most of his career as a ranch and rodeo horse in addition to use as a stallion. The Peakes purchased him in 1945. Some of his better-known get were Filaree, Idleen, Guinea Pig II, Show Boy, Fox, Red Risk, Silver Lad, and Glider. His daughters, bred to Driftwood or his foals from mares by King, Waggoner, or Popcorn, produced outstanding results for Rancho Jabali. The well-known Coaster was by Tony out of Rock Sea by King, and bred at Rancho Jabali. A number of Tony daughters were bred to Speedy Peake, by Driftwood, to produce foals like Brandy Peake, Scotch Peake, Drift-Along Katy, Northwood, and Sequoia Wood. Road Runner P, a Driftwood daughter, produced Jabilina P when bred to Tony.

Fig. 5-2. Photo courtesy Peake files.

El Greco 2227 was a good looking sorrel son of King P-234 foaled in 1940, and out of Old Sugar by Foreman by Elmendort TB, a U. S. Remount stallion. He was bred by Jess Hankins of Rock Springs, Texas. His foals, out of both Driftwood and Tony daughters, helped set the standard toward which Rancho Jabali was reaching.

Quicksand 8270, a 1944 chestnut son of Popcorn by Red Dog and out of Ruby Wear by Tony, was utilized by both Rancho Jabali and Perry Cotton. He combined the Casement bloodlines of Ballymooney, by Concho Colonel, with the Traveler line through Tony. Quicksand was a show winner and good rodeo mount in addition to being a successful sire. He stood at the Cotton Ranch in Visalia.

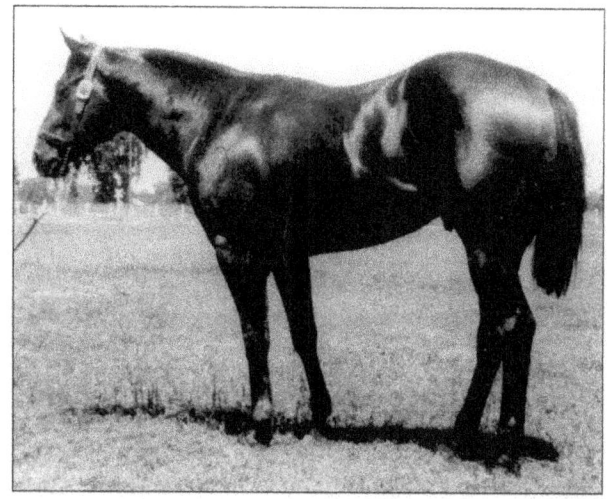
Fig. 5-3. Photo courtesy Peake files.

Sun Raider 2871, by Popcorn and out of Raggedy Ann by Tony. He carried on the Casement/Possum combination of blood lines, producing performance horses of more blocky conformation than many of the Driftwoods.

Fig. 5-4. Photo courtesy Peake files.

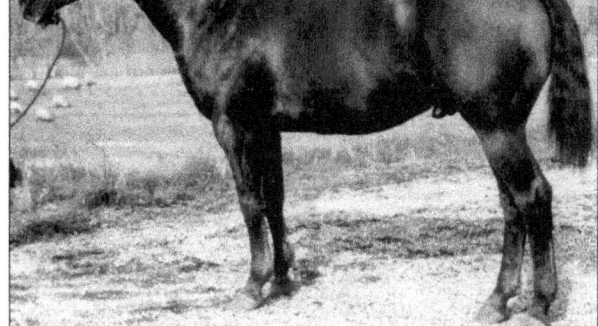

Booger H. P-12901, a 1945 chestnut son of King and out of Queen by Dan by Old Joe Bailey (Weatherford). He was a successful race, rope, and show horse in his own right before being retired to stud. Booger H. was owned, campaigned, and bred by Perry Cotton. Both he and El Greco passed on the King influence in the Rancho Jabali and Perry Cotton horses.

Fig. 5-5. Photo courtesy Peake files.

Fig. 5-6. Photo courtesy Peake files.

War Chief 279, by Joe Hancock and out of a Burnett mare, was acquired as an outcross on the Driftwood daughters to combine the speed and proven performance of Joe Hancock and Driftwood. He was eventually given to Tex Oliver who used him for the rest of his life. Bred to Brown Beulah, by Driftwood, he sired War Drift and War Concho, both stallions who have had an impact on the usin' horse world.

Tex Oliver was one of the first breeders to recognize the capabilities of the Driftwood/Joe Hancock cross through War Chief and utilized it very effectively throughout his long and successful career as a breeder. Oliver knew what type of horse he wanted to raise and planned several generations ahead to achieve his ideal.

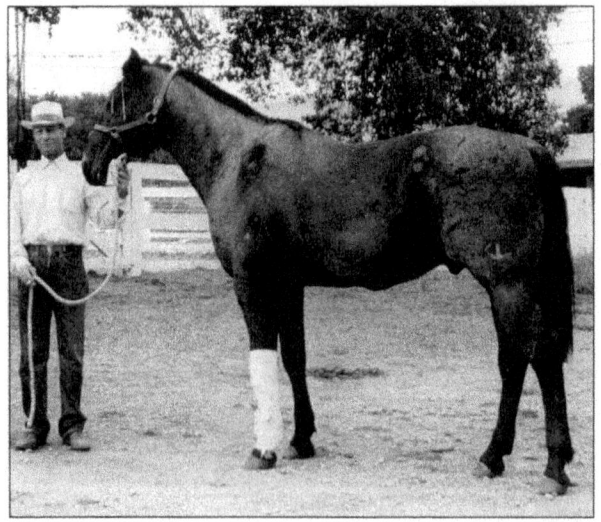

Fig. 5-7. Photo courtesy Peake files.

Roan Hancock 456 upon his arrival at Rancho Jabali. The 1935 red roan stallion was by Joe Hancock out of a Burnett mare. His conformation was typical Hancock—big, heavy-boned, and very functional. The Driftwood-Hancock cross provided some of the top usin' horses of the time. Many of his early sons became outstanding ranch and rodeo horses in the Southwest and his daughters were good producing broodmares. Injured in a trailer accident, Roan Hancock died at Rancho Jabali after siring only two or three foals. If he had lived, the foals he might have sired out of Driftwood daughters would probably have been exceptional performers under saddle.

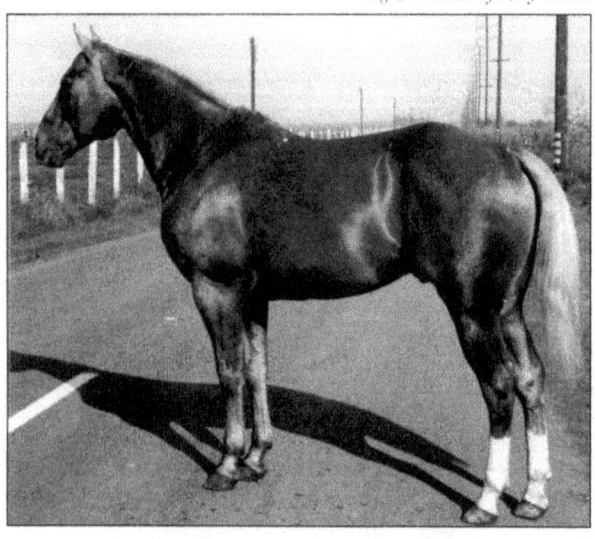

Fig. 5-8. Photo by Perry Cotton.

Braz d'Or P-2679 This 1940 palomino son of San Simeon/Lady Coolidge was leased from Perry Cotton and used as a stallion. Braz was a top rope horse and later a successful cutting horse. A number of his daughters were crossed with Driftwood and produced rope, show, and ranch horses. Braz d'Or was later purchased by Cal Poly, where he stood for a number of years. When the four Driftwood daughters given to Cal Poly by the Peakes were bred to Braz d'Or the results were impressive.

6

THE RANCHO JABALI MARES

"...that was the best load of mares that ever left the Hughes Ranch and I hope that they make the same reputation in California that they have in Texas."
—Duwain Hughes

RO mares
(Greene Cattle Company, Patagonia, Arizona)
Hornet
Nosey
Waspy
Dos X

Hughes mares
Sage Hen
Jack Rabbit
Prairie Dog
Pigeon
Smoky McCue

King mares
Shu Cat
High Tone
Spiderette
Queen Anne

Other mares
Nugget Hug
Hancock Belle
Straw
Hildegund
Trixie
Clara Bow M
Keeta 7
Dusky Ruth
Sweet Hallie

While Driftwood was the major factor contributing to the success of Rancho Jabali, the broodmares were the base on which it was built. Those individuals, all of solid Quarter Horse conformation and foundation bloodlines, produced foals that found a strong demand among buyers.

Catherine Peake was careful to utilize mares which generally carried more Quarter Horse characteristics than did Driftwood. When the King- and Hancock-bred mares were crossed with him, this produced a stouter horse with more bone. Driftwood passed on speed, athletic ability, an adaptable mind, and the alert "Driftwood look" which came to characterize his offspring.

Katy was also strong in her resolve not to breed Driftwood to excitable, unbroken, or unproven mares. She felt that inherited disposition, coupled with the influence a quiet mare had on the foal while it was nursing, would carry on down. An outstanding performance horse had to have the disposition and willing mind to accept training.

While she produced a number of excellent horses by Driftwood, Katy was not blind to the possibilities of different stallions. Many of the Rancho Jabali mares were bred to stallions other than Driftwood to see what different combinations would produce.

Katy had the "horseman's eye" to an exceptional degree and was able to visualize what a cross would probably look like as an adult animal. This gave the Rancho Jabali program a focus which few other breeders have been able to match.

Fig. 6-1. Photo courtesy Peake files.

RO mares at Rancho Jabali, 1944. *Left to right: Waspy, Dos X, Wood Chuck, and Dusty Miller.*

Hornet 2190, sorrel, foaled in 1939 by El Rey RO 896 out of an RO Quarter mare. By Driftwood she foaled:
　Stinger B
　Driftwood Monty.

Nosey 2192, sorrel, foaled in 1939 by El Rey RO 896 out of an RO Quarter mare. She was the dam of:
　Little Speed
　Speedy N.

Waspy 2196, sorrel, foaled in 1940 by El Rey RO 896 out of an RO Quarter mare. By Driftwood she produced:
　Wood Wasp,
　Driftwasp
　Waspwood.

Dos X 2517, sorrel, foaled in 1937 by El Rey RO 896 out of an RO Quarter mare. By Driftwood she produced:
　Tres X,
　California Sweetheart,
　Jabalina P
　Zoe Driftwood.

Fig. 6-2. Photo courtesy Peake files.

Sage Hen 2194, roan, foaled in 1939 by Waggoner 2691 out of a Harmon Baker mare. Katy Peake described her: "Sage Hen was a clever moving mare, a little waspish, and a natural born boss in the pasture. She was a wonderful and protective mother and a tremendous milker." Her foals included:
　Straw 2195 by Nevada Bomber No. 916,
　Drifty P-6668 by Driftwood,
　Road Runner P. P-8848 by Driftwood,
　Red Button P-10781 by Driftwood,
　Sagy by Yellow Jacket,
　Blue Dog by Tony,
　Watch Dog by Tony,
　Henny Penny Peake P-22534 by Driftwood,
　Poker Chip Peake P-22536 by Driftwood,
　a red roan horse colt by Driftwood,
　a red roan filly by Bob Cat,
　Drifting Sage by Driftwood
　Roanie Boy by Driftwood.

Pigeon P-11958, roan, foaled in 1939 by Waggoner 2691 out of an Anson mare by Harmon Baker. Her only Driftwood foal was:
   Cherokee Jake.

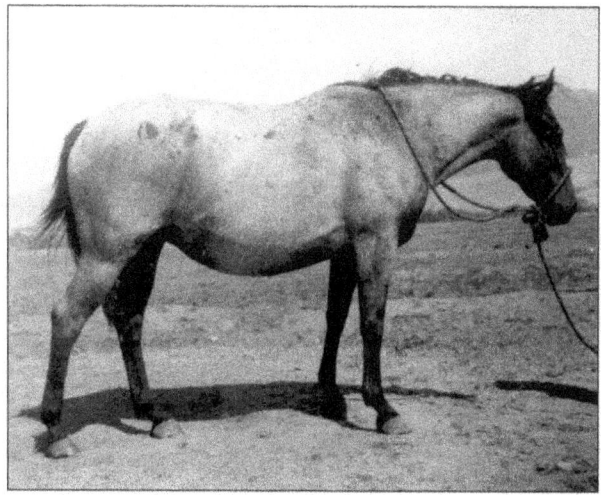

Fig. 6-3. Photo courtesy Peake files.

Smoky McCue 2518, gray, foaled in 1939 by Waggoner 2691 out of a Harmon Baker mare. By Driftwood she produced:
   Dogwood,
   Mac McCue W,
   Easy Keeper,
   Smoke Maker
   Drift Easy.

Fig. 6-4. Photo courtesy Peake files.

Jack Rabbit 2191 *(photo not available)*, roan, foaled in 1939 by Waggoner 2691 out of a Harmon Baker mare. By Driftwood she produced:
   Jima Red.

Prairie Dog 2193 *(photo not available)*, roan, foaled in 1939 by Waggoner 2691 out of a Harmon Baker mare. With Driftwood she produced:
   Scooter Mac
   Katy Red.

Fig. 6-5. Photo courtesy Peake files.

Shu Cat P-5619, black, foaled 1943 by King P-234 out of a Hankins Quarter mare. Her Driftwood foals included:
- **Driftwood II,**
- **Peakewood,**
- **Speedy Scat,**
- **Speedy Peake,**
- **Kittywood**
- **Runaway Peake.**

Fig. 6-6. Photo courtesy Peake files.

High Tone P-2189, sorrel, foaled in 1942 by King P-234 out of Uncle's Pet 2505. Her only Driftwood foal was:
- **Fleet Wood.**

Fig. 6-7. Photo courtesy Peake files.

Spiderette P-13359, black, foaled in 1942 by King P-234 out of Spider H. P-13102 by Darity TB. Her Driftwood foals were:
- **Swiftwood,**
- **Jernigan Peake**
- **Wood Mite.**

Queen Anne 2781, chestnut, foaled in 1942 by King P-234 out of a Holland mare by a son of Harmon Baker. By Driftwood she foaled:
　Miss Linwood,
　Brown Beulah
　Annie Wood.

Fig. 6-8. Photo courtesy Peake files.

Rock Sea P-3232, by King and out of Silly by Show Boy. She was bred to Tony, producing:
　Coaster.

Fig. 6-9. Photo courtesy Peake files.

Nugget Hug P-8456, sorrel, 1944, by Bear Hug 2868 out of Goldienug 8451 and bred by Sid Vail of Douglas, Arizona. By Driftwood she produced:
　Hug Me Tight,
　Quick A Lick
　Wooden Nugget.

Fig. 6-10. Photo courtesy Peake files.

Fig. 6-11. Photo courtesy Peake files.

Hancock Belle P-5593, dun, foaled in 1944 by Buck by Chicaro out of Miss Hancock by Joe Hancock. Her Driftwood foals included:
  Buckwood,
  Maestro,
  Chilena,
  Driftwood Ike
  Speedywood.

Fig. 6-12. Photo courtesy Peake files.

Dusky Ruth 34067, sorrel, foaled in 1949 by Lucky Blanton out of Bajita Fifi by Hard Tack. When bred to Driftwood she foaled:
  Dusky Peake,
  Speedy II,
  Speedy Darnell
  Woodmist.

Fig. 6-13. Photo courtesy Peake files.

Sweet Hallie 2689, bred by Jack Casement by Red Dog out of Buckeye. She foaled:
  Chipperwood 55344 by Driftwood.
  Chipperwood was first owned by Bordon Chase, then by Roy Patton, and finally by John Hoyt, where he sired a number of Working Cow Horses that set the show ring afire.

Straw 2195 *(photo not available)*, roan, foaled in 1942 by Bomber 916 out of Sage Hen. She foaled:
  Blue Glory by Driftwood.

Hildegund P-4886 *(photo not available)*, sorrel, foaled in 1943 by Tony out of Frisky Wear by Tony. When bred to Driftwood she produced:
  Greasewood
  Northwood.

Trixie *(photo not available)*, an unregistered mare by Ace owned by Joe Stephens of Matador, Texas. When bred to Driftwood she produced four daughters which were registered with the AQHA. They were:
  Clara Bow M 9452, 1938;
  Nellie S 9363, 1939;
  Rosy G 9453, 1942;
  Flickers S 9455, 1943.

Keeta 7 8271 *(photo not available)*, sorrel, foaled in 1945, by B Seven by Balmy L out of Jo-Mo-Quita by Pal-O-Mine, bred by O. W. Caldwell. By Driftwood she foaled:
  Keetawood,
  Firewood,
  Woodkee,
  Woodfern
  Just Seven.

A number of other mares were utilized in the Rancho Jabali program, bred to either Driftwood or one of the other stallions. Photographs of the Rancho Jabali mares show a uniformity of conformation, sufficient bone and traditional Quarter Horse characteristics. There are, naturally, some family differences, but overall they are the type which would produce performance horses.

While the produce of Driftwood sons was limited (many were gelded and went to the rodeo or show arenas and ranches, or were not bred extensively), his daughters, granddaughters, and great-granddaughters made substantial contributions to the performance horse world. A close look at the extended pedigrees of many outstanding usin' horses will find the name of Driftwood linked with one of the mares listed above. For well over half a century their influence has continued.

Fig. 6-14. Photo courtesy Peake files.

Tony daughters used by Rancho Jabali. Left: Honey Bug out of Shu Cat. When bred to Driftwood she produced Bugaboo Peake. Right: Jamboree out of Kitty Rose.

Fig. 6-15. Photo courtesy Peake files.

Lowry Girl 114 50252, by Roan Hancock P-456 and out of Lowry Girl 65 by Charlie Hart P-9245. By Driftwood she produced Travelog 81494 and Buttons Cumming 97884.

One example would be **Miss Linwood 32525.** By Driftwood and out of Queen Ann by King, she produced six outstanding foals by six different stallions. These were:
   Cibecue Roan by Red Man;
   Jake Stewart by Mr. Hancock;
   Pillo by Revenue H.;
   Vee R Ironwood by Bond Issue;
   an unnamed sorrel stallion by Kaweah Twister;
   One Nite Stand by Clown Hancock, an outstanding broodmare in her own right.

Swiftwood, another Driftwood daughter, produced the sensational working horse and sire:
   Red Wood Man by Red Man.

Fig. 6-16. Photo courtesy Fred and Clara Wilson.

Miss Linwood 32525, *by Driftwood and out of Queen Ann by King. This prolific producer of athletic talent displays the typical "Driftwood look" which has made the strain so recognizable over the years. She produced seven foals by seven different stallions, including Cibecue Roan by Red Man.*

An inspection of the Rancho Jabali breeding and foaling records from 1942 through 1963 shows that various mares were bred to different stallions to find which combination "nicked" best. It is also interesting to survey the horsemen who brought outside mares to the ranch and which stallion they bred to. Included on the list are Chet Behan, Al Lauer, Dale Smith, Joe Yanez, Harlan Brown, Buck Nichols, Ronnie Russell, Walter S. Markham, Roy Wales, Tom Mattart, Chuck Sheppard, Gordon Davis, Will Gill, Jr., Hap Magee, Buck Nichols, Tex Oliver, Perry Cotton, Ralph Camarillo, Raymond Cornelius, Joe DeYong, Bordon Chase, Bill Adair, Bill Lamkin, Kemper Marley, and Glen Buell.

A number of those individuals contracted for the foal out of a certain Rancho Jabali mare at breeding time. A look at the breeding and sale records confirms "repeat orders" over the years. Once a usin' horse man rode a Driftwood, he wanted another.

Oklahoma Gal (Buckskin Belle), by Double Star by Oklahoma Star and out of Hancock Belle. By Driftwood she foaled:
  Oak Wood.

Fig. 6-17. Photo courtesy Peake files.

Tiny Tot Peake 174629, daughter of Driftwood and Balmy Gal by Balmy L with her Easy Bar colt, and in foal to Braz d'Or. Before becoming a broodmare, Tiny Tot Peake was a good calf roping mount. She was eventually sold to Suzanne Koch of Stevinson, California, and was the basis of her successful Appaloosa program.

Fig. 6-18. Photo courtesy Peake files.

Farewell P-2397, by Lucky Blanton and out of Betsy R by Red Cloud. She was not bred to Driftwood.

Fig. 6-19. Photo courtesy Peake files.

Fig. 6-20. Photo courtesy Fred and Clara Wilson.

One Night Stand and her foal, Drifter Heels Two. This good Driftwood granddaughter, by Clown Hancock and out of Miss Linwood (Driftwood), was bred by Jake Kittle and owned by Fred and Clara Wilson of Newcastle, Wyoming. Producers such as this mare have kept the Driftwood legacy alive and well. As a broodmare she has produced:

Clownwood by Clown Two;
Driftin' Heels by Senator Man;
Driftin' Heels Two by Senator Man;
Maybe Wood by No Maybes;
Maybe Driftin by No Maybes,
7-time barrel racing qualifier to the PRCA Badlands Circuit Finals and twice to the National Circuit Finals Rodeo;
Maybe Cottonwood by No Maybes,
Open Superior Barrel Racing and Open Performance ROM;
Woodsy Maybes by No Maybes;
Nother Time by No Maybes;
Driftwood Maybes by No Maybes,
team roping qualifier for the PRCA Badlands Circuit Finals Rodeo;
Hot Drift by Doc's Compact Model.

Fig. 6-21. Photo courtesy Peake files.

Lucky Lady Tucker 34821, by Lucky Blanton and out of Patches. Bred to Driftwood she produced:
Tweedle Dum.

7

DRIFTWOOD'S GET WHO MADE THEIR MARK

"He...left behind a legacy of diverse accomplishments among his descendants which continues to proliferate."

—Katy Peake

Buckwood - *1948*
Chakaty - *1954*
Cherokee Jake - *1944*
Chilina - *1950*
Chipperwood - *1955*
Cowboy Schell - *1935*
Cowboy C - *1940*
Drift O Smoke - *1964*
Driftwood II - *1950*
Driftwood Ike - *1954*
Driftwood Maid - *1963*
Drifty - *1943*
Dusky Peake - *1953*
Dusty Driftwood - *1961*
Easy Keeper - *1945*
Firewood - *1952*
Henny Penny Peake - *1949*
Jernigan Peake - *1954*
Jake - *unregistered*
Mac McCue - *1943*
Maple Wood - *1963*
Mescal Brownie - *1955*
Pepperwood - *1960*
Poker Chip Peake - *1950*
Red Button (Roany) - *1945*
Road Runner P. - *1944*
Roany Boy - *1953*
Speedy II - *1955*
Speedy N - *1945*
Speedy Peake - *1956*
Speedywood - *1956*
Tom Skidwell - *unregistered*
Wooden Nugget - *1957*
Wood Wasp - *1946*
Woodwind (Hot Toddy) - *unregistered*
Wood Tick - *1949*

Driftwood's contribution to the performance horse world is substantial. His blood is still sought because of the achievements of his get, grandget and extended family—that "they performed" and a rodeo or stock horse contestant "could win on them." Without that continued usability under saddle, proving that he would "breed on," Driftwood's name would probably only be a footnote in the annals of the American Quarter Horse.

Cowboy Schell No. 16612   Cowboy Schell, a blaze-faced bay gelding that stood 14.3 and weighed 1035, was foaled in 1935 at Silverton, Texas. Raised by the McClouds, his dam was a Roy McMurtry Quarter mare by Will Steed by Billy McCue. Cowboy first appeared in Arizona as a cow horse on the Boquillas Land and Cattle Company, coming from Texas with Wayne McCloud. According to Johnny Burson, Cowboy carried McCloud the whole way from Texas to Arizona instead of riding in a trailer. The Texan knew what he had in the speed department and occasionally matched the bay gelding—who carried on the family tradition of getting across the finish line first. It wasn't long before Cowboy went to the track instead of gathering cattle.

Page Lee saw Cowboy race against Clabber and Lucky Blanton. Although the gelding was beaten by those two speedsters, Lee was impressed. He purchased Cowboy in 1937. A short while later, Cowboy outran Lucky Blanton at Benson, Arizona. His time for the 440 was :22.4 (unofficial).

Until 1945, all Cowboy did was race. He proved to be a top sprinter at any distance up to the quarter. At the Hacienda Moltacqua Track (near Tucson) Cowboy was listed for 300 yards in :16 2/5; 350 yards in :18 3/5; and 440 yards in :23 1/5. In 1943 he crowded Clabber to track record time for 350 yards and finished ahead of Red Man and Painted Joe. He defeated the celebrated Arizona Girl as well as Red Racer in a match race for

*Cowboy, a 1935 bay gelding by Driftwood (front), noses out **Red Racer** in a quarter mile match race at Hacienda Moltacqua. The time was :23 3/5 seconds. At the time he was owned by Page Lee of Benson, Arizona.*

***Cowboy** puts **Asbury Schell** in roping position.*

the quarter in :23 3/5. He also outran such noted sprinters as Idleen, Texas Lad, Golden Slippers and Rumpus.

In the 1945 edition of the Southern Arizona Horse Breeders yearbook (following his registration with the Quarter Horse Association), he was awarded the designation of Celebrated American Quarter Horse as well as a Register of Merit.

In 1945 Page Lee sold Cowboy to Abe Graham of Chandler as an arena prospect. The bay gelding was getting a little old for continued track work. Graham, a top hand with both a horse and rope, trained Cowboy for calf and team roping as well as bulldogging.

That fall Buckshot Sorrells and Homer Pettigrew, two of the best timed event hands in professional rodeo, took Cowboy to New York City for the annual rodeo at Madison Square Garden. Cowboy was still "high" from the race track but did well over the short score and with the big cattle that were used at the Garden.

A few months later, in 1946 at Scottsdale, Asbury Schell saw Cowboy. He had sold Driftwood to Katy and Channing Peake and was looking for a replacement. The left-handed Cliff Whatley was riding Cowboy, although Graham owned him. Asbury was winning the calf roping with Whatley still to make his run.

"When I came up," remembered Whatley in an article by Willard H. Porter, "Cowboy threw a whing-ding in the roping box. He was used to the short score at New York and wanted to break the instant he saw the calf. We were roping over a long score at Scottsdale. When the calf came out of the chute, I had to hold the horse back to keep from breaking the barrier."

"Well, Cowboy reared up and turned all the way around with me," continued Whatley, "and the calf was really going to water. After I straightened him out and started after it he felt like the fastest horse I'd ever been on. I caught the calf and managed to tie it, just barely beating Asbury out of the money. Asbury said afterwards that he wanted any horse that could do that—and he soon got him."

Schell paid Abe Graham $3,500 for the bay gelding. In the first two rodeos after the purchase, the horse had more money won off his back than Schell had paid for him.

At Lehi, Utah, Asbury won first in the calf roping average and third in the team roping. Eddie Schell (his son), riding Cowboy, and Maynerd Gayler took the team roping. Gayler, on Cowboy, placed second in the calf roping. The second rodeo was Reno, Nevada. Joe Bassett, of Tonto Basin, Arizona, was heeling off of Cowboy and topped the team roping with Pete Grubb.

During the ten years that Asbury Schell hauled the bay gelding, he only raced him once. Matched against the well-known Juanita at Durango, Colorado, he nosed out the mare for 200 yards.

His "big lick" was the rodeo arena and money was won off of his back in calf roping, dally team roping, team tying and bulldogging. Frequently, he carried a number of different riders at a contest and always got them to cattle in money-winning time.

Cowboy C 8502 This 1940 bay son, out of Mickey by Cimarron, was registered with the AQHA as being bred by Joe Bassett of Tonto Basin, Arizona and owned by George Cline, also of Tonto Basin. Based on the foaling date, Driftwood was owned the year before by Cline so Bassett took advantage of his neighbor's good stallion. George liked the resulting colt so well that he purchased him. Cowboy C followed his sire's example and became a top calf and steer roping mount for both George and his brother Leck. At one time, Leck Cline was heeling behind Asbury Schell when he was riding Cowboy Schell, making a pair of Driftwoods that often brought home team roping purses.

As a rodeo mount, Cowboy C was ridden in both the calf roping and team roping by George and his brother Leck. Old-time ropers still remember how the bay stallion could catch cattle over a long score and the way his hard stop made good calves out of big, rank, bad ones. As one rodeo hand of the period recollected, "The Clines were tough, and they had most of us out-mounted."

Cowboy C was not used as a stallion extensively but he did produce ranch and rodeo horses for Arizona cowboys. The majority were probably not registered with the AQHA since "papers" were not that important at the time for a working gelding. A daughter, Cookie Driftwood, out of Ranger Cookie by Old Red Buck, was bred to Doc O'Lena. She produced Doc O'Lenawood who foaled Peppy Pizazz by Peponita.

Drifty P-6668 This brown stallion (later gelded) was the first of several outstanding horses that resulted from mating Driftwood to Sage Hen. According to Channing Peake, "Other than a small white star, Drifty looked exactly like his dad. He matured into a horse that wasn't quite as tall as

Fig. 7-3. Photo courtesy Lexie Lobapesky.

Cowboy C *also served as a queen mount for several years at the Payson rodeo. Three of the girls who rode him won.*

Driftwood, but heavier—taking after his dam in that respect." The liaison was arranged when Schell and Driftwood visited Ranch Jabali in 1942. Drifty was foaled in 1943 and Asbury purchased him as a weanling, feeling that he needed "another Speedie." The colt followed his sire to the rodeo arena. Asbury's son, Eddie, heeled on the horse for several years before selling him to Tommie Rhodes, two-time PRCA World Champion Steer Roper.

Under the ownership of Tommie Rhodes, a Mammoth, Arizona rancher, Drifty was an outstanding team roping mount (he was one of the pair of horses which took the team tying at Phoenix one year). During his long career he helped ropers collect checks all over the Southwest. His performance helped establish the reputation of the Driftwood horses as top contest mounts.

He was also used as a sire to a limited degree. While many of his sons and daughters were not registered with the AQHA, they did carry on the Driftwood legacy and made good contest and ranch mounts for Arizona cowboys.

Road Runner P P8848 Second in the Driftwood/Sage Hen line was the roan mare Road Runner P. Used as a broodmare, Road Runner P produced twelve foals by different stallions, including Jabalena P; Finito, Open ROM; and Ebb Tide, who earned a Racing ROM.

Red Button (Roany) P-10781 Third of the nick between Driftwood and Sage Hen was the roan Red Button foaled in 1945. On the rodeo circuit he was better known as "Roany." Sold as a yearling to Pat Smith of Hermiston, Oregon, Roany rapidly developed into the kind of mount a man could "win on."

Smith was roping calves off of the stout colt at two and on the professional rodeo circuit with him a year later. With his quick speed and hard stop, Roany was an ideal mount for the Northwestern rodeos where the scores were long and the calves big. He was also an excellent heeling horse, both in dally roping and team tying. In fact, the roan was so accomplished that Bill Linderman once offered $3,500 for him.

Roadrunner P 8848 by Driftwood and out of Sage Hen. This good mare was the dam of Red Runner; Finito, Open ROM; Miss Pockett; Real Cat; Road Show; Run Requested; Miss Casbar; Buddy Jim; Jabali Squaw; Ebb Tide, Racing ROM; and Miss Front Page.

Like many of the Driftwoods, Roany was also a tough match racer and his owner was ready to back his horse. Because of his rodeo experience, Roany was easy to score and would walk up to the line with no fuss, breaking when he got the signal and running straight for the finish line. And he was never headed.

As a stallion, Red Button sired only a limited number of foals. Smith was more interested in roping on Roany than standing him at stud. Roany was gelded in later life.

Red Button (Roany), by Driftwood and out of Sage Hen, comes to a stop for *Pat Smith* at the Ellensburg, Washington rodeo.

In 1952 the roan horse changed ownership. Demose Berevin of Pendleton was searching for a mount for his 16-year-old son, Joe. He figured that Roany would be the best one he could get on. It took some tradin'—and 11 head of Black Angus cows to boot—to get the job done. Joe and the horse "fit," winning lots of calf, team roping and team tying money for his owner in Oregon, Washington and Idaho arenas.

Henny Penny Peake P-22534 The fourth product of the Driftwood/Sage Hen combination made her reputation in the show ring. Henny Penny Peake, a 1949 bay mare, aptly demonstrated the athletic ability that was her birthright. She brought fans to their feet when she settled into a sliding stop or went down the fence with a cow. Under the capable handling of Jimmy Williams, the talented mare won the Pacific Coast Hackamore Championship in both 1953 and '54—then went on to show everyone how it was done in the Bridle, competing in Reined Cow Horse classes. During her show career she earned the prestigious titles of both AHS and PCHJSHA Stock Horse of the Year.

As a broodmare, Henny Penny Peake produced the ROM Jimmy Wood, Henny Penny Poo, show winner Speedy Penny, Silky Peake and Red Pepper Wood.

Henny Penny Peake also brought national attention to the Driftwoods when she appeared as the star of the Walt Disney feature film "The Bay Lady." She and

Jimmy Williams puts the fabulous **Henny Penny Peake** through her paces in the hackamore at a California show during the 1950s. The mare showed the typical Driftwood looks, refinement and athletic ability.

beautiful Sammy Fancher made quite a team on the screen.

Sammy Fancher remembered the mare as "A little on the 'hot' side but an outstanding performance horse and a pleasure to ride."

Poker Chip Peake P-22536 To the calf ropers of the 1950s and '60s, Poker was considered the greatest rope horse ever to step into the arena. The 1950 gray gelding, fifth in the winning Driftwood/Sage Hen combination, could catch cattle, stop so hard that he stood big, rank calves on their heads and always worked a perfect rope.

Oscar Walls of Tempe, Arizona, bought Poker Chip from Rancho Jabali as a weanling. By the time that the skittish colt was two, Walls was roping calves and heading and heeling steers on him. His next owner was Duane Ellsworth of Chandler. Ellsworth had J. K. "Red" Harris rope on the colt. He eventually let Harris have a half-interest in him. Later, Harris purchased the other half.

J. K. knew that the nervous gray had the potential to be a great rope horse, but he had to settle down. In order to accomplish that the roper spent two years "making" the horse. In addition to roping at home, he trailered to rodeos and rode him around, letting his spooky pupil get used to the noise, the strange sights and the confusion. Occasionally when he had drawn a good straight-running calf, he would rope off the gray. That strategy worked and Poker Chip began to fulfill the promise that Harris had seen in him.

The last man to own the sensational gray was Dale Smith of Chandler. J. K. had decided to hang up his ropes, so in January of 1955, Smith forked over a sizable chunk of cash for the smooth working gelding. Surprised at the price tag, Dale's father remarked, "With all the horses we have, why do you need to spend that much money on another one?" Several years later, after he had seen how his son was winning on the spectacular gray, he advised against taking a $10,000 offer from Willard Combs.

Dale Smith and Poker Chip Peake went on to make rodeo history. Smith was twice PRCA World Champion Team Roper, a Calf Roping top-fifteen contestant numerous times, the first man to qualify for the NFR in three events (Calf Roping, Team Roping, Steer Roping) and competed at the NFR a total of twenty times.

In January 1956 Smith took Poker Chip to his first indoor rodeo at Denver. He was the first roper, tied his calf in 19 flat and the horse worked perfectly. On the second calf, Poker Chip stopped so hard and got back on the rope so quickly that the animal wouldn't get up, costing the roper his chance at both a day money and the average. Observing the action, Calf roping legend Toots Mansfield remarked, "Dale, your horse cost you the day money. But I wouldn't worry—I've been trying all my life to get a horse to work that way."

From then on, the gray gelding was on the road, building his reputation as "the best calf horse in the business." During his career, money was won off of his back at every major rodeo, the National Finals and numerous matched and jackpot ropings. While Smith didn't make a practice of mounting lots of ropers on Poker, most of the "greats" had a seat on the horse at one time or another—and frequently collected a check.

Fig. 7-7. Photo courtesy <u>Western Livestock Journal</u>

Dale Smith roping on Poker Chip Peake, *a well-known son of Driftwood, at the big Salinas, California rodeo. Poker's stop in this picture is "textbook perfect."*

In addition to his performance under saddle, Poker Chip was easily recognized by his smooth conformation and sparkling grey color. He gained a host of fans, not only among ropers but spectators as well. The first National Finals Rodeo, held at Dallas, Texas in 1962, was televised and Poker Chip was a standout.

In 1959 Dale roped at San Francisco's Cow Palace rodeo which was televised. Upon his return home, neighbor Roy Wales greeted him with, "Poker looked great on television. He made Roy Rogers' Trigger look like an old dawg."

In January of 1963 Poker Chip was injured. On the way to Denver from Odessa, Texas, the trailer turned over when the driver swerved to miss a cow in the road. Poker and the other horse were pinned in their stalls. The gray gelding was badly bruised on his back and withers as well as the hind legs.

The injured Poker Chip was taken to Dale's brother Max, a veterinarian. For several months he was rested and treated. Then he went back into competition. He worked well except that his speed was curtailed because of an injured hind leg.

Finally, as a last resort, Smith took his horse to the University of California School of Veterinary Medicine. There, Dr. John Wheat (the same veterinarian who treated Poker's half brother, Speedy Peake) operated on the injured leg. He peeled away the calcium deposit and bad tissues caused by a blood clot in the area of the injured muscle. The operation was a success, but Poker still wasn't himself.

Feeling that rodeoing would be too hard, Dale retired Poker Chip to the home place at Chandler. One visitor, noting the plush quarters the horse enjoyed, commented that it looked as if the horse had it made. Smith's answer was, "He can sleep in the house if he wants to. He paid for it."

Mel Potter of Marana, Arizona won on the horse a number of times and commented, "He was like a magic wand." At Red Bluff, California one year he was mounted on Poker Chip and really filled his wallet. "We were roping big, rank Angus calves, with the cows at the back end. They could really run and were tough to handle. Dale was first in both go rounds and the average and I was second in both goes and the average. Poker really made the difference for us."

Poker Chip Peake was gelded fairly early. His projected performance as a rope horse probably had a bearing on that. He did sire a few foals. Two of them were Poker Pot and Poker Player. The authors have found little information on them so they must have been gelded and followed their sire to the roping arena. There was also a daughter who became the dam of Missle Bar.

Poker Chip Peake is buried on the Trail of Great Cow Ponies at the National Cowboy Hall of Fame in Oklahoma City, Oklahoma. Frequently, an old-time roper is seen pausing in front of the grave in memory of one of the greatest horses ever to chase a calf down a rodeo arena.

Drifting Sage 44426 Foaled in 1950, this blue stallion went to the Northwest where he became not only a solid rope horse but a successful sire as well. Because he was not campaigned heavily he didn't receive the notoriety as his siblings—Drifty, Road Runner P, Red Button (Roany), Henny Penny Peake, and Poker Chip Peake. He did produce a number of good offspring, including Blue Chip Driftwood and Blantonwood, and his name is found in many pedigrees today. He was later gelded.

Roanie Boy The last of the Driftwood/Sage Hen foals was the blue roan gelding Roanie Boy. An outstanding ranch and team roping horse, he was never campaigned like his brothers Poker Chip or Red Button, and not well known outside of California.

Jernigan Peake 48030 This brown gelding, foaled in 1954 out of Spiderette P-13359 by King, is considered to be one of the greatest reined cow horses to set foot in the show ring. Bred by Perry Cotton and owned by the Peakes, Jernigan Peake was sold to Bill Danna who had John Brazil Jr. train and show him. He campaigned for several years, winning all of the big contests on the Pacific Coast. In 1961 he was California Champion Reined Cowhorse. Jernigan Peake was undefeated in the Hackamore, competing against the top horses of his day, including Johnny Tivio. In the Bridle Jernigan Peake won such shows as the Cow Palace, Salinas, Monterey and Sacramento. Those contests were either American Horse Show Association or California Hunter-Jumper/Reined Cow Horse Association approved. Consequently, he has no AQHA points to his credit.

John Brazil, Jr. wrote, "He was perhaps the greatest horse that I had the pleasure of riding. He had speed, cow and athletic ability, plus the conformation and bone structure that we all desire in a horse."

Bridle horse fans still talk about how the brown gelding could stop and turn around, then dominate a cow during the cattle portion of the stock horse event. The late Greg Ward, one of the finest trainers and competitors of the period, considered Jernigan Peake one of the top five stock horses of all time.

Driftwood II P-22532 This brown stallion out of Shu Cat by King was foaled in 1950. He was a full brother to the versatile Speedy Peake. Owned by Al Lauer, and later by Albert and Annette Carlton of Woodside, California, Driftwood II accumulated AQHA points in Halter, Barrel Racing, Western Riding, Working Cowhorse, Reining, and Western Pleasure, earning an Open Performance ROM in 1957. He later broke his leg, which curtailed his ability under saddle although he was a successful sire.

Easy Keeper P-12044 A bay son of Driftwood/Smoky McCue foaled in 1945, Easy Keeper was purchased at the 1946 Rancho Jabali sale for Will Gill, Jr. by Ray Cornelius of Santa Ynez. The yearling colt had no name, although he had been nicknamed "Apple" by Katy Peake because of his color and great hindquarters.

Easy Keeper was started under saddle as a two-year-old. He was later turned over to Floyd Boss, a great hackamore and cutting horse man, for fine tuning. Gill took the young stallion home to the ranch and he spent the rest of his life there. Used in the feedlot, the bay stallion never let a steer get by him at the sorting gate and could really make his rider "screw down." Out in the pastures, he could "walk up a storm"—a trait he passed on to most of his get. Not only would he watch a cow, but like his sire, he could run and stop. Will Gill, Jr. roped

Fig. 7-8. Photo courtesy Peake files.

Driftwood II. *He was originally selected by Katy Peake as a replacement for Driftwood. Sold to Al Lauer, he was a winner at both halter and performance.*

Fig. 7-9. Photo courtesy Will Gill, Jr.

Easy Keeper, *by Driftwood and out of Smoky McCue. He was owned by Will Gill, Jr. of Madera, California, and sired a host of good ranch and roping horses.*

Fig. 7-10. Photo courtesy Peake files.

Speedy Peake, *by Driftwood and out of Shu Cat, was selected by Katy Peake to become the senior stallion at Rancho Jabali.*

calves and showed Easy Keeper at a few nearby Quarter Horse shows until the stallion was injured in a breeding accident and retired.

While Will Gill, Jr. was primarily concerned with raising ranch horses, numerous Easy Keeper sons went to the rodeo arena. Twice, the stallion won the trophy for having the most colts ridden at the big California Team Roping Championship at Oakdale. One of his grandsons won the Outstanding Head Horse award in 1995. Jim Rodriquez, Jr. roped on Gill (registration records lost), an Easy Keeper son, for 11 years, winning the PRCA World Champion Team Roper title four times. On Gill, Rodriquez also won both the Oakdale 10-steer and Chowchilla 8-steer ropings. Cadillac, registered as Easy Gran and out of a Red Mud mare, was Easy Keeper's last son. Foaled in 1974, the year Easy Keeper died, Cadillac took Jim Petersen and David Gill to the NFR in 1985. Jake Barnes rode him in the last three go-rounds to cinch the first of his seven World Championships in the Team Roping. Other top headers who have had a seat on the big gelding include World Champions Tee Woolman and David Motes and Top Fifteen ropers Don Gatz and Jim Wheatley. His last owner was Joe Murry of Oakdale, California.

Easy Keeper sired a total of 185 foals for the Gill Ranch. He helped establish the reputation of the Gill horses as some of the best in the nation.

Speedy Peake P-62167 - ROM in 5 events Foaled in 1956 out of Shu Cat by King, Speedy Peake was selected by Katy Peake to carry the flag. He had it all—breeding, looks, athletic ability and a mind which let him excel in different events.

Trained by Max Schott, Speedy Peake demonstrated that he was a willing pupil. Schott campaigned him in Cutting, Working Cow Horse, Calf Roping and Team Roping. Many times, Speedy Peake would place in Cutting, then place again in Working Cow Horse and finish up with points earned in both Calf and Team Roping. In 1965 the brown stallion won the Working Horse Stake at the prestigious Santa Barbara show. Like most of the Driftwoods, Speedy Peake was a great calf horse who scored well, got to cattle quickly, had a terrific stop and worked a rope. In the opinion of many ropers, he would have been a winner on the pro-rodeo circuit.

In the spring of 1962, it looked as if the stallion's performance career was over. Playing in his paddock, Speedy Peake kicked through the wire fencing, hung up and dislocated a back leg, as well as breaking the hock. Max Schott called his veterinarian, Dr. Vern Reitan, who splinted the leg.

Upon Dr. Reitan's recommendation, Speedy Peake was taken to the Veterinary Hospital at the University of California. There, he was operated on by Dr. John Wheat.

Speedy Peake remained at Davis for several weeks. Although immobilized in a sling, he proved to be an excellent patient. Early X-rays showed that the bone had knitted and the stallion was taken back to the ranch. Dr. Reitan prescribed turning him out in a well-fenced paddock to wait and see. In the spring of 1963 Speedy Peake was X-rayed sound and started back under saddle. Max Schott said, "We took it slow and easy."

In 1964 Speedy Peake was shown only four times. At the Cow Palace (San Francisco) he won the Steer Stopping and was third in the Senior Registered Cutting. At King City he won one go-round in the Calf Roping (fastest calf roped during the show) and second in the average. He was also second in Novice Cutting, second in Steer Stopping and third in Team Roping. He won the Calf Roping at the Santa Barbara Fiesta and was third in the $1,000 Stallion Stake at Santa Maria.

He was shown only four times during 1965, winning the Santa Ynez Valley Horse Show, the Tri-County Working Stock Horse Trophy Saddle, and the Team Roping at the Santa Barbara Fiesta. He also won the $1,000 Championship Working Horse Stake at Santa Barbara.

Fig. 7-11. Photo courtesy Peake files.

Sparrow Peake, *by Speedy Peake and out of a One Eyed Waggoner mare. She was the Hackamore Champion at the San Francisco Cow Palace one year.*

Remembering the last win Max Schott chuckled, "There were five events in the Santa Barbara Stake. We placed first in Cutting, Calf Roping and Steer Stopping, and second in Working Cow Horse. Then there was a Pleasure class and we placed fifth in that... heh, heh, heh."

In recalling Speedy Peake, and Max Schott's talent to bring out the best in the stallion, Perry Cotton remarked, "Max was a hand. He liked the horse, worked with him and had the ability to develop his performance to its fullest."

Following the dispersal of the Rancho Jabali horses in 1966, Speedy Peake was leased to Barbara J. Lonstreth of Modesto, California. He died in 1968 from colitus X, with his last crop of foals hitting the ground in 1969.

In ten foal crops, Speedy Peake sired 113 registered foals with 24 winning points in AQHA shows. Many of his sons were campaigned as reined cow horses in the open shows on the Pacific Coast (where their winnings did not show up in AQHA listings) or as rope horses. His daughters went into broodmare bands where they passed on his good looks and athletic ability.

Speedy Peake did carry on Driftwood's heritage, not only as a performer himself but as a sire of performers. In doing so, he justified Katy Peake's faith in him.

AQHA Get Performance Summary
SPEEDY PEAKE

Open Halter Point Earners	4
Open Halter Points	7
Open Working Point Earners	18
Open Working Points	68
Open Registers of Merit	3
Open Performers	21
Amateur Performers	1
Youth Working Point Earners	4
Youth Working Points	13
Youth Performers	6

Wooden Nugget P-96078 One of the more versatile sons of Driftwood, this 1957 bay stallion out of Nugget Hug by Bear Hug (dam of triple AAA horses Nug Bar, Easter Bee, and Miss Wonder Bar), was bred by Roy Wales. He was later purchased by the Peakes.

Trained and exhibited for Rancho Jabali by Walt Mason, Wooden Nugget excelled in working cow horse and stock horse classes on the California circuit. Shown 15 times in 1964 and '65, the bay garnered seven firsts, winning three trophy saddles; four seconds; one third; and three fourths. In AQHA competition he earned points in Western Pleasure.

In the show ring, "Woody" beat such top horses as Night Mist, Johnny Tivio, Fiddle D'Or, Sonny Joe Reed, and Mr. Grey Ghost.

Returning to Roy Wales after the Rancho Jabali dispersal in 1966, Woody became a professional quality calf roping horse. Loaned to B. J. Pierce for the summer rodeos, he was a winner at Pendleton, Ellensburg, Walla Walla, Pueblo, and other Northwest contests where the scores were long and the calves rank. At the big Wilcox, Arizona matched calf roping, Pierce averaged 11 seconds on four calves to take first money against the toughest ropers in the business.

After the roping Dale Smith asked B. J. Pierce, "Is that a Wales horse?"

B. J. simply nodded his head.

"That figures," replied Smith.

Turned out in a wheat pasture to rest after the rodeo season, Wooden Nugget was crippled. He was never used as an arena mount after that. The production record on the stallion is limited so he must have been bred to only a few mares. Wooden Nugget's last AQHA-recorded owner was the Miller Livestock Company in Livingston, Montana.

Dusky Peake P-41535 This 1953 son of Driftwood was out of Dusky Ruth by Lucky Blanton, combining the bloodlines of two proven rope horses. Dusky Peake was a solidly built bay standing 14 hands 3 1/2 inches and weighing 1,175 pounds, making him "just the right size" for a contest horse. He was owned and roped on by John

Fig. 7-12. Photo courtesy Peake files.

Wooden Nugget *by Driftwood and out of Nugget Hug by Bear Hug. With* **Walt Mason** *riding, "Woody" was an outstanding working cow horse—then went on to be a pro-caliber calf roping mount.*

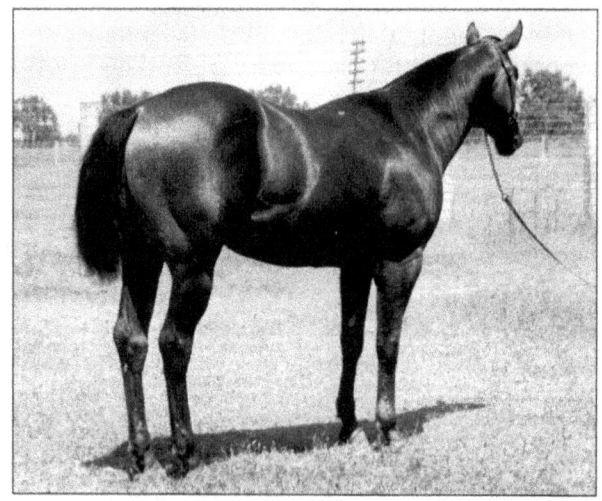

Fig. 7-13. Photo courtesy John Bacon.

Dusky Peake, *a son of Driftwood and out of Dusky Ruth by Lucky Blanton. Dusky Peake was owned and roped off of by John Bacon.*

Bacon of Los Olivos, California. Like many of the Driftwood sons, he was not bred extensively, but his foals could all do the job under saddle. They ended up working on California ranches or in the roping arenas.

Buckwood P-22531 This 1948 dun mare was out of Hancock Belle. Owned by Al Lauer, she was the 1961 Pacific Coast Trail Horse Champion and National American Horse Show Association

Woodwind and Jimmy Williams. This bay mare was one of the great hackamore and bridle horses of the 1950s and '60s.

High-Point Horse. Later she became a professional rodeo quality heel horse under Frank Ferreira before being retired as a broodmare.

Woodwind (Hot Toddy) Out of the Thoroughbred Easter Gale, she was another top bridle horse that won her share at contests up and down the Pacific Coast under the capable handling of the "old master," Jimmy Williams. Owned by Mr. & Mrs. Al Lauer of Sacramento, California, and competing against the great Shady Lady, Hooker D and the Arabian Ronteza, Woodwind earned the 1961 title of Pacific Coast Stock Horse Champion. Since she was not registered with the AQHA and did not compete in AQHA shows, her winnings were at ASHA- and CHJRC-sanctioned shows.

Sammy Fancher, who rode both Woodwind and Henny Penny Peake on occasion, said that she preferred the former. The mare was much more easy going than Henny Penny Peake, although she had the same athletic talent. "Woodwind was a pleasure to ride," remembered Fancher.

Wood Wasp P-19955 This 1946 brown mare was out of Waspy P2196 by El Rey RO 896. She was owned by W. P. Roduner of Merced, California and was trained as a cutting horse. Her ability in the arena placed her in the Top 10 Pacific Coast Cutting Horse Association standings in 1952. As a broodmare she produced Miss Woodwasp and Waspy Girl.

Driftwood Ike, by Driftwood and out of Hancock Belle. Owned first by Roy Wales of Queen Creek, Arizona, and then by the K4 Ranch of Prescott, Arizona, Ike was an outstanding calf and team roping mount in addition to being a top sire of rope horses. His get and grandget are still winning in the arena.

Driftwood Ike 47645 - ROM A smoky dun stallion out of Hancock Belle, Driftwood Ike was foaled in 1954. That made him a full brother to Buckwood, Chilina, and Speedywood. Purchased by Roy Wales of Queen Creek, Arizona as a two-year-old, "Ike" was introduced to the arena and rapidly developed into an outstanding calf roping and team roping mount.

In the breeding pen Ike has been considered Driftwood's most influential son.

Walt Nichols took Ike to the Prescott PRCA rodeo when he was just three, placed in both go-rounds, and won the average in the calf roping. That set the tone and many of the name ropers would get a seat on Ike when they were in Arizona. Gordon Davis, Sonny Davis, B.J. Pierce, Chuck Sheppard, Walt and Don Nichols, Roy and Jess Patton, Bill Rush, Bill Lowe, John Miller and Abe Logue all competed on Driftwood Ike at one time or another —and frequently took home a check.

Chuck Sheppard, a tough calf and team roper (a World Champion in the latter event) remembered the first time that he saw Driftwood Ike. "It was at the Clovis, New Mexico rodeo and B. J. Pierce won the calf roping on him. I really liked him." In 1963, at Tucson, Arizona, Sonny Davis won the calf roping on him and Sheppard, who also had a seat on the smoky dun stallion, split 3rd and 4th.

Wales' oldest daughter, Jerry, ran barrels on the versatile stallion at both horse shows and rodeos, being out of the money only one time. That was usually at the same contests where one or more people were roping on Ike. The change of events, or riders, never seemed to make any difference to him.

One of the things that Jerry Wales Sanford remembered about Driftwood Ike was his extremely calm disposition. She wrote, "He never got in a hurry or excited. The only time that he hurried was from the time he was backed into the roping box until your hands went up to signal for the flag. Then, it would take five minutes to ride him out of the arena."

Ms. Sanford continued, "Ike looked at everything when ridden down the road or out on the desert— he could never just walk. He would always gawk at this and that and his tracks would look like a drunk person."

In the show arena, Driftwood Ike won calf and team roping points, earning his Open Performance ROM in 1960 and 2nd High Point Steer Roping Horse in 1969. One year at the big Phoenix show, Driftwood Ike and Firewood (both Wales horses sired by Driftwood) were first and second in each of the two go-rounds and tied for the average in the calf roping. Driftwood Ike won a total of 54 AQHA Performance points in roping events.

In 1964, John Kieckhefer and Chuck Sheppard purchased Driftwood Ike. He would stay at the K4 Ranch in Skull Valley, Arizona for the rest of his life.

Fig. 7-16. Photo courtesy Wales Ranch.

Roy Wales on Firewood *by Driftwood and out of Keeta 7. Wales was a long-time fan of Driftwood, owing Driftwood Ike, Firewood, and Wooden Nugget in addition to raising a herd of foals by them.*

AQHA Get Performance Summary
DRIFTWOOD IKE

Total Registered Foals	215
Crops	25
Race Starters	8
Starts	53
Wins	8
Winnings	$8,699
Racing Registers of Merit	3
Leading Money Winner	Nifty Ike
Halter Points	14.0
Total Performance Points	236.5
Performance Point Earners	33
Performance Registers of Merit	5
AQHA Champions	1

Well-known breeder Mel Potter has built his program around both daughters of Driftwood Ike and a son, Lone Drifter. Stanley Johnson, another influential Driftwood breeder, owned and used Orphan Drift for many years.

As a sire, Driftwood Ike produced 215 AQHA-registered foals, including numerous outstanding usin' horses as well as name stallions and broodmares. Some are Wayward Ike out of Katy Mas; Orphan Drift out of Orphan Annie C; Lone Drifter out of Moore Yen; Ikes Last; Tuffy Ike; and White Lightning Ike who has sired so many good horses for Will Gill and Sons. He also sired Nifty Ike, a track record holder and one helluva rope horse; Driftwood Lady, AQHA Champion; Heart Wood, 5th Place AQHA High Point Steer Roping Horse; and Flutterby Katy, racing ROM. His name is in the pedigrees of many top rodeo horses today. Rodeo Champions such as Clay O'Brian Cooper, Allen Bach, Mike Beers, Brad Smith and Tom Ferguson are all roping on, or have roped on, sons and grandsons of Driftwood Ike.

Firewood 38362 Nicknamed "Dobie" by the Wales family, out of Keeta 7 by Balmy L., this bay gelding was foaled in 1952. He was not only a competent rope horse, earning his Open Performance ROM in 1957, but a halter winner as well. At Prescott one year he won his halter class—then came back and won the calf roping that afternoon.

Bred by Katy Peake and owned by Roy Wales of Queen Creek, Arizona, Firewood and his stablemate Driftwood Ike dominated the AQHA roping contests in the Sunshine State during their show careers. And like Driftwood Ike, Firewood was a popular mount with professional ropers when they were in the area. A contestant could win roping calves and heading or heeling steers on the versatile gelding.

Chipperwood 55344 This brown stallion, foaled in 1955, was out of the Red Dog daughter Sweet Hallie, and combined two performance dynasties. Bred by Bordon Chase, a Hollywood screen writer with a love of good horses and western history, Chipperwood was sold to Arizonan Roy Patten for a rope horse. From Patten he went to the well-known John Hoyt of Arizona. While he did not accumulate a show record of his own, his foals made quite an impact in the Working Cow Horse, Reining and even Working Hunter classes.

Chipperwood was used as a stallion to a limited degree, siring only 44 foals in 9 crops. However, twelve of those foals went to the show ring and distinguished themselves as horses to beat by amassing an impressive 163.5 points in Open competition and 284.5 in Youth for a total of 456.

Drift O Smoke 401766, out of Neta B Waggoner 153429. This 1964 roan mare, with John Hoyt in the saddle, collected 54.5 Performance points plus an Open Performance ROM in 1967 and Open High Point Working Cow Horse in 1969. Drift O Smoke ended up in California carrying a junior rider to multiple wins in the Calforina Reined Cow Horse Association.

Maple Wood 302676, out of Dusty Apple 77184. This versatile 1963 dun mare collected 28 Open points as well as 4 in the Youth division. Among her accomplishments was a 1972 Open Performance ROM and a 5th in the 1972 Youth division of the AQHA World Show in Working Hunter Under Saddle.

Driftwood Maid 245638, out of Judy Joy 77462. This 1963 buckskin mare was truly a capable performer. She amassed 30 Open Performance points and 150 Youth Performance points, plus a 1971 Youth Open ROM, 1974 Open Performance

---

AQHA Get Performance Summary
CHIPPERWOOD

Halter Points 8
Halter Point Earners 3
Open Performance Points 163.5
Youth Performance Points 284.5
Total Performance Points 456.0
Performance Point Earners 8
Performance Registers of Merit 5
Point Earners in All Divisions 10

ROM, as well as 3rd in the 1975 Youth World Show Trail Horse and in 1976 Youth Superior Western Pleasure.

Dusty Driftwood 231119, out of Cow Gal Hancock 121650. This brown 1961 mare collected an Open Performance ROM in 1966, a Youth Performance ROM in 1967, and earned 31 Open Performance points and 122.5 Youth Performance points, for a total of 153.5.

Pepperwood 146175, out of My Valentine 43377. This 1960 chestnut mare accumulated 70 Open Performance points during her career.

While the Chipperwoods were not numerous, they did carry on the Driftwood legacy of performance. Their record at AQHA shows was impressive.

Mac McCue W 6751 Known as "Old Blue" among Arizona ropers, the 1943 grey gelding out of Smoky McCue did his share to carry on the Driftwood heritage. Owned by Oscar Walls of Tempe, Arizona, Old Blue was a great roping horse, who was equally adept on either calves or steers and earned an ROM in 1957. Like his sire, Mac McCue S could run. His sleepy eye and easy-going attitude disguised his sensational speed for a short distance, which made him easy to match. Walls ran the horse 57 times, winning 50 of them. Oscar Walls once sold the gray for $3,500—and then bought him back.

Chakaty 48005 Bred by Buck Nichols, this 1954 daughter of Driftwood/Lady Lux N (Lucky/Toots by Wave of Eire TB) became a top calf roping and heeling horse for John Fincher and Don Nichols. Under Fincher, who trained her, she successfully competed during the middle 1960s before turning her over to Nichols (he was going on a 2-year LDS mission). In 1960, John and Don "went partners" and traded Chakaty to Al Lauer for Chilina, another Driftwood daughter. While Lauer owned her, Chakaty earned the 1961 Pacific Coast Champion Western Pleasure Horse title, demonstrating the adaptability of the Driftwoods. She was remembered as a mare of class and unique action.

As a broodmare, Chakaty produced seven good foals: Arctic Katy, Toddy Wood, Chatty Wood, Chakaty Wood, Katchina Wood, Katy Blue, and Wooden Sheik.

Fig. 7-17. Photo courtesy Peake files.

Mac McCue S 6751, *by Driftwood and out of Smoky McCue, pictured as a two-year-old before he turned gray.*

Fig. 7-18. Photo courtesy John Fincher.

Chakaty *as a two-year old. She was foaled in 1954. The brown mare was by Driftwood and out of Lady Lux N.*

Chilina 34066 A solidly built, smoky dun 1950 mare was out of Hancock Belle, making her a full sister to Driftwood Ike, Buckwood and Speedywood. Katy Peake sold her to Don Nichols as a weanling. Chilina got her start in the arena as a calf and heel horse, winning her first roping as just a two-year-old. Nichols sold Chilina to Al Lauer, then, in partnership with John Fincher, traded back for her. (Nichols put up $2,500 and Fincher put up Chakaty, another Driftwood daughter.) John later bought out Nichols and rodeoed on the mare for several years before trading her to Bob Ellsworth for a little bunch of Angus heifers. Chilina went on to produce six foals: Redwood Beaver, Stetson

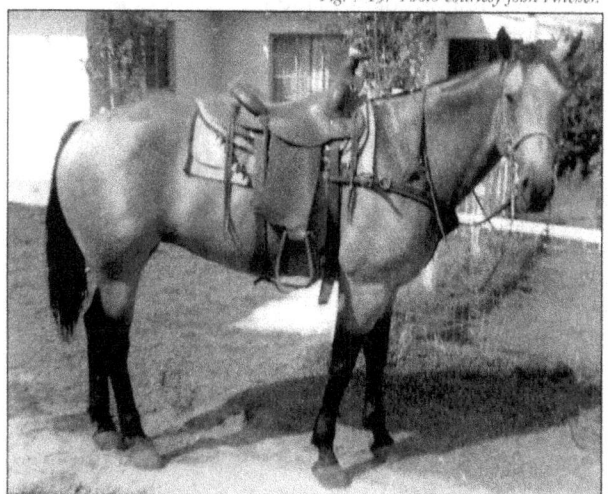

Fig. 7-19. Photo courtesy John Fincher.

Chilina. *This 1950 daughter of Driftwood/Hancock Belle was a very successful calf roping and heeling horse for both John Fincher and Don Nichols before becoming a good broodmare.*

Fig. 7-21. Photo by Willard Porter.

Cherokee Jake *with* **Asbury Schell** *at a rodeo.*

Eddie Schell on Cherokee Jake *and* **Dale Smith on Poker Chip Peake** *at the 1955 Salinas rodeo. These two sons of Driftwood won lots of roping money for their riders.*

Bar, Miss Chilina, Mr. Outside, Frostens, and Lena's Mack.

Cherokee Jake 7366 This roan stallion was foaled in 1944 out of Pigeon by Waggoner. He was bred by Rancho Jabali, purchased by Jake Stine of Santa Barbara, California, and later sold to Asbury Schell. Under Schell he did double duty as both a ranch and rodeo mount. He was considered one of the stoutest horses ever on the end of a rope in the team tying event and could really pull a big steer down after Asbury had heeled it. Cherokee Jake eventually ended up with John Clem, another tough Arizona roper.

Cherokee Jake did sire a number of usin horses and broodmares for Arizona horsemen before he was gelded. Perhaps his best known son was Shady, owned and ridden by Dale Smith. Aboard the black gelding, Smith garnered two World Champion Team Roping titles as well as roping calves and tripping steers. Smith was the first man to qualify for the National Finals Rodeo in three events (Calf Roping, Team Roping, and Single Steer Roping), working the last two events on Shady.

Speedy II 65035 A 1955 brown stallion, out of Dusky Ruth by Lucky Blanton, spent his life in Arizona under the saddle of Sam McKinney. He was an outstanding head horse who, according to John Fincher, "could get across the line to cattle as fast as any horse I ever saw."

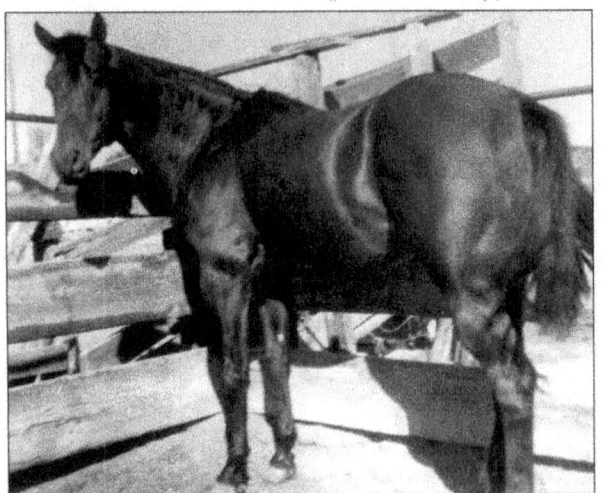

Fig. 7-22. Photo courtesy John Fincher.

Speedy II, *a 1955 son of Driftwood/Dusky Ruth by Lucky Blanton. Owned by Sam McKenny, Speedy II was a top-rated head horse at Arizona rodeos and ropings in addition to being a good sire.*

He was a full brother to Dusky Peake, Speedy Darnell, and Wood Mist. Since he was in Arizona, the Nichols bred a number of Clabber mares to him.

Mescal Brownie 113135 A 1955 brown gelding out of Lazy Mark 38245 by Mark distinguished himself under Chuck Sheppard. He earned his Open ROM in 1960, then went on to carry his owner to the pay window at numerous pro rodeos across the Southwest in both calf and team roping. Sheppard, a former World Champion Team Roper, heeled on Brownie at the 1963 PRCA National Finals Rodeo in Los Angeles.

Speedywood 56971 This 1956 brown stallion, by Driftwood and out of Hancock Belle, was a full brother to Driftwood Ike, Chilina, Maestro, and Buckwood. Bred by Catherine Peake, Speedywood was sold to Al and Norma Lauer in California. Eventually he went to Dale Smith of Chandler. At the time Smith purchased the stallion he was told that there was a broken bone in his foot and he would never be sound enough to ride. Speedywood was turned out with Smith's mares for two years, ranging over the rocks and cactus in the rugged pastures—and came back sound.

Not one to let a good horse go to waste, Dale Smith began roping calves and headed and heeled steers on the stallion when he wasn't busy in the

Fig. 7-23. Photo courtesy Western Livestock Journal.

Mescal Brownie, *by Driftwood and out of Lazy Marke, winning the heeling at Tucson, Arizona. With **Chuck Sheppard** in the saddle. Mescal Brownie won lots of rodeo and jackpot money, including making an appearance at the 1964 NFR at Los Angeles, California.*

Fig. 7-24. Photo courtesy Dale Smith.

Speedywood, *a 1956 son of Driftwood out of Hancock Belle. An open ROM winner in 1966. Speedywood was first owned by Al Lauer of California who showed him, then by Dale Smith of Chandler, Arizona who roped on him and bred the stallion, and finally by R. E. Odem of Orange, Texas. Speedywood is shown with owner **Dale Smith** after a successful calf roping run.*

Fig. 7-25. Photo courtesy John Fincher.

John Fincher *pitches his slack away and prepares to step off of* **Tom Skidwell**, *an unregistered son of Driftwood, at the 1963 Showlow, Arizona rodeo.*

Fig. 7-27. Photo courtesy Max Schott.

Three of Speedy Peake's foals and their sire *at Rancho Jabali. From left, the riders are Walt Mason; unknown; Elwyn Hall; and Max Schott on Speedy Peake.*

Fig. 7-26. Photo courtesy Joe Murray.

Vern Castro *roping on his Driftwood gelding,* **Jake,** *at the Calgary Stampede during the early 1950s. This horse was later owned by Dr. Frank Santas.*

breeding pen. In addition to Smith, Eddie Schell, Mel Potter and Billy Hamilton all competed on the capable stallion. Speedywood sired a number of good horses including the stallion Speedyman, Reg Camarillo's Wildfire, Julio Moreno's Six Pac, Emm Dee owned by Pete Zanettie and daughters who produced such performers as Tom Ferguson's Bar. Speedywood's last owner was R. E. Odem of Orange, Texas.

Tom Skidwell   An unregistered 1958 son of Driftwood out of Mary Jane, a daughter of Red Man and an RO mare. Like Driftwood, Mary Jane was a pro-caliber calf horse, taking Del Haverty to many wins before being purchased by Katy Peake. Tom Skidwell was bred by Katy Peake and purchased by John Fincher as a five-year-old. The bay gelding became a money-winning calf roping mount and carried his owner to an Inter-collegiate Calf Roping title in the 1960s.

Jake   Among Driftwood's sons who made a name in the rodeo arena was the unregistered Jake. Owned and campaigned at first by Vern Castro of Castro Valley, California (1942 RCA co-Champion Team Roper with his brother Vic, and 1949 runner-up), and then by Dr. Frank Santos, the bay gelding did double duty as both a calf roping mount and a head horse. Over the long roping scores used at

Fig. 7-28. Photo courtesy Tex Oliver.

Wood Tick P-26909 *by Driftwood and out of Miss Honda. Bay, 1949.*

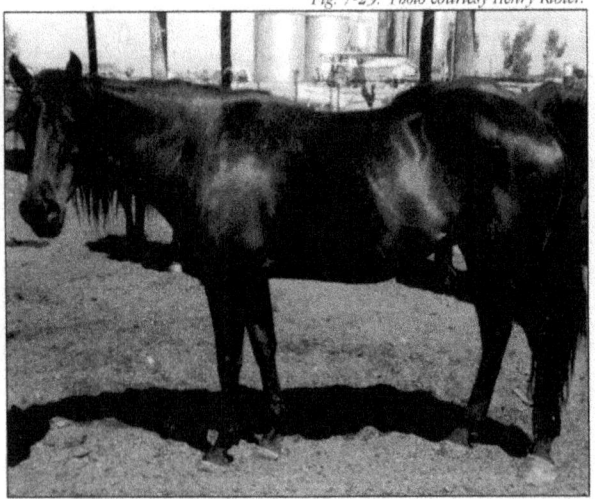

Fig. 7-29. Photo courtesy Henry Kibler.

Brown Beulah 38833, *a 1952 Driftwood daughter out of Queen Ann by King. She was bred by Perry Cotton and owned by Tex Oliver for many years. Her last owner was Henry Kibler. Her foals by different stallions were Vee R Peggy Ann, Drifting May, Wood Peake, Beulah's Annie, War Drift, War Concho, Mahala Hancock, Woody Brown, Medulce, Maybelline, and Redwood Jake.*

Fig. 7-30. Photo courtesy Peake files.

Speedy N 17045, *foaled in 1945, by Driftwood and out of Nosey.*

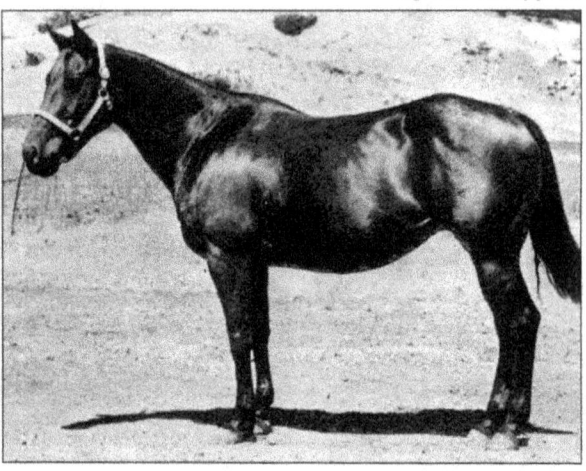

Fig. 7-31. Cal Poly photo.

Anniewood, *by Driftwood and out of Queen Ann. Anniewood was one of the mares presented to Cal Poly by Katy Peake in 1957.*

Pacific Coast rodeos, Jake proved his ability to catch cattle in a hurry and then turn big, rank calves into good ones. At one time or another, the capable gelding carried various riders to winnings at all of the major rodeos in the area. Like most of the Driftwoods, his sturdy conformation and good feet and legs kept him going for many years.

Santos, who purchased the gelding from Castro, won the All Around Champion Cowboy at the Cow Palace and at Salinas on Jake. Since Frank Santos was a practicing veterinarian he didn't compete as regularly as most ropers did, but when he went, he was capable of winning. He once remarked that "Jake was the best horse I ever rode."

Wood Tick 26909 This brown horse, a 1949 son of Driftwood out of Miss Honda by Hank by King, was bred by W. G. "Tex" Oliver of Buellton, California. Oliver was one of the early Driftwood fans, and during his long career as a breeder raised a number of outstanding horses. He knew what he wanted in a horse, was aware of the blood combinations that it took to produce that horse, and was able to plan years and generations ahead to get it. Wood Tick was Oliver's personal ranch mount and spent his life cowboying on the different ranches where Tex was employed.

Speedy N 17045 This 1945 bay gelding out of Nosey by El Rey RO was bred by Catherine Peake. He was purchased at an early age by Parr Norton, a rancher and roper from Gateway, Oregon. Beyond those facts, little is known about the horse, although he must have continued the Driftwood tradition of "making a cow horse."

8

DRIFTWOODS AND THE TOUGHS WHO RODE THEM

"They made good horses."

—*Munroe Tumlinson*

Driftwood
- Buck Nichols
- George Cline
- Leck Cline
- Asbury Schell
- Eddie Schell
- Roy Wales

Cowboy Schell
- Ab Graham
- Asbury Schell
- Eddie Schell
- Maynerd Gayler
- Buckshot Sorrells
- Homer Pedigrew

Smokey
- Chester January
- "Red" Deussen
- Billy Joe Deussen
- George Seals
- Bobby Seals
- other North Texas ropers during the 1940s and '50s

Goober
- Bill Bently
- Hubert Briggs
- George Seals
- other North Texas ropers during the 1940s and '50s

Snake
- Wylie Ennis
- numerous other North Texas ropers

(continued)

As a Quarter Horse family, the Driftwoods have undoubtably carried more rodeo cowboys to a paycheck than any other within the breed. While it is impossible to determine just how much money has been won off of their capable backs, both in jackpot ropings and rodeos, there is no doubt that it totals well into the millions of dollars. At the professional level (PRCA rodeos), numerous world Championships have been earned by contestants riding a Driftwood-bred horse.

The ropers who have been lucky enough to ride the Driftwoods have stories about them. In each case, the story covers how well the horse worked under unusual circumstances and what "naturals" they were.

Munroe Tumlison of Cresson, Texas is a life-long roper. He has qualified for the National Finals in both the Calf Roping and Single Steer Roping, as well as training horses which other ropers rode and won on.

A number of years ago he trained several sons of Speedywood, by Driftwood, for R. E. Odem as calf roping mounts. "They made good horses," he remembered.

One black gelding (Tumlinson doesn't remember his name) was on the snorty side, an alert colt that could run and was quick on his feet. That stood him in good stead and he would really stop and work a rope. Once he learned that when the roper's hand went past his head with the slack rope, it meant stop, "he'd really fall against the ground!"

"It took a long time but, when it did come, he made one hell of a horse," smiled the cowboy.

The first rodeo Tumlinson took the horse to was Cheyenne,

Poker Chip Peake
  Dale Smith
  Oscar Walls
  "Red" Harris,
  Lanham Riley
  John Clem
  Herschel Romaine
  Lee Cockrell
  Sonny Davis
  Billy Hamilton
  Eldon Dudley
  Lee Ferris
  Mel Potter
Mac McCue W (Old Blue)
  Oscar Walls
  Gilbert Nichols
  John Clem
Chipperwood
  Roy Patton
  Jess Patton
  John Hoyt
Drifty
  Asbury Schell
  Eddie Schell
  Tommy Rhodes
Cherokee Jake
  Asbury Schell
  Eddie Schell
  John Clem
Smoke Maker
  Will Gill, Jr.
  (he was riding this horse when
  he won the big Oakdale, CA
  roping with Buck Jones)
Scooter Mac
  Lefty McPeters
Buckwood
  Al Lauer
  Frank Ferriera
Speedy Peake
  Max Schott
Wooden Nugget
  Walt Mason
  Roy Wales
  B. J. Pierce
Cowboy C
  George Cline
  Leck Cline

Driftwood Ike
  Roy Wales
  B. J. Pierce
  Chuck Sheppard
  Sonny Davis
  Gordon Davis
  Walt Nichols
  Don Nichols
  Bill Rush
  Bill Lowe
  John Miller
  Stan Harder
Firewood
  Roy Wales
  B. J. Pierce
Bill Adair
  Charlie Mickle
Henny Penny Peake
  Al Lauer
  Jimmy Williams
  Sammy Fancher
Wood Wind (Hot Toddy)
  Jimmy Williams
Wood Wasp
  W. P. Rodner
Redwood Jake
  Jake Kittle
  John Hoyt
War Drift
(out of a Driftwood mare)
  Tex Oliver
Doc's Drift
(out of a Driftwood mare)
Orphan Drift
  Stanley Johnson
Speedywood
  Al Lauer
  Dale Smith
  Mel Potter
  Eddie Schell
  R. C. Odem
Double Drift
  "Doc" Whitman
  Freeland Thorson
Jaywood
  Kathryn Donovan
Red Button (Roany)
  Pat Smith
  Joe Bergevin
Easy Keeper
  Will Gill, Jr

Driftwood II
  Al Lauer
  Ray Yanez
Maestro
  Eddie Schell
Jernigan Peake
  Bill Dana
  John Brazil, Jr.
Haliwood
  Roy Wales
  Abe Logue
Chilina
  John Fincher
  Don Nichols
  Gilbert Nichols
  Al Lauer
Buckwood
  Al Lauer
  Frank Ferriera
Speedy II
  Sam McKinney
Driftwood Peake
  Ralph Camarillo
  Leo Camarillo
  Jerold Camarillo
Jake H
  Tommy Rhoades
Dusky Peake
  John Bacon
Shady
(by Cherokee Jake)
  Dale Smith
  Mel Potter
  Homer Pedigrew
Mescal Brownie
  Chuck Sheppard
Wood Wind
  Jimmy Williams
Button Wood
  Jimmy Williams
Peako Peake
  Don Dodge
Speedy Cat
  Bill Gibford
Chakaty
  John Fincher
  Don Nichols
  Al Lauer
Tom Skidwell
  John Fincher

(continued)

which was a testimonial to his confidence in the animal. Monroe qualified for the final go-round in the 15th hole and drew a runner for his last calf.

"It was over the long 30-foot score and we were really rolling when I pulled up a-straddle of the calf halfway down that big arena. I reached out, roped him, and the bottom fell out the way that black horse stopped. I tied the calf in 16 seconds and moved from 15th to 9th in the average—and they were only paying eight monies."

As Tumlinson rode out of the arena, Tuffy Thompson's wife told him that he was riding the best horse at Cheyenne.

"He ought to be," replied Monroe. "It's his first rodeo."

John Fincher, now of Plattsmouth, Nebraska, was another roper who "did pretty good" on the Driftwoods. In 1964, rodeoing for Arizona State University in Tempe, Arizona John won the Intercollegiate Calf Roping title. He was riding Skidwell, an unregistered bay gelding by Driftwood and out of a Red Man mare. He purchased the horse from Rancho Jabali in 1963 as a 5-year-old and made a calf horse out of him.

"At Roy Wales' urging, I went to Rancho Jabali and looked at two geldings," remembered Fincher. All they'd done was just ride Skidwell but I figured he'd work because of the way he was bred. Katy Peake was asking $2,000 for him. When I told her that all I had was $1,500, she took it. I won a lot on the horse."

Raised in Arizona, John grew up around Buck Nichols and learned the value of a good horse. He also got his start as a roper in the Nichols' pen, associating with many of the "toughs" of the time like Roy Wales, Jess Patton, John Clem, the Ellsworths, and the Arnolds. As a teenager, he started both Woodfern and Chakaty (daughters of Driftwood) under saddle for Nichols (1956) and rode both of them to school. "It was seven or eight miles to school and I'd ride one of them one day and the other the next. I'd tie the horse up during the day, water it at noon and then ride back home in the evening. I purchased Chakaty from Buck for $500, as a two-year-old, after I'd ridden her to school."

Fig. 8-1. Photo courtesy John Rudnick.

*Seven-time PRCA World Champion Heeler **Clay O'Brian Cooper** riding **Casper** by Call Me Ike out of OH April by Driftwood Ike. Another PRCA Champion Heeler, Mike Beers, also rode Casper during his career.*

Nichols sold Woodfern to Stanley Johnson and he bred her to Poco Speedy. The resulting foal was Qui Qui, the mother of Wilywood, Tom Eliason's Driftwood stallion that is now 24 years old (2001). Woodfern was Wilywood's maternal granddam and his sire was Orphan Drift, making him a double-bred Driftwood.

In the summer of 1958, the summer after John's freshman year at Arizona State University, the young

Fig. 8-2. Photo by Wagner; courtesy John Rudnick.

*Many-times PRCA World Champion Team Ropers **Leo** (heeling) and **Jerold Camarillo** (heading) doing what they do best—rope. Jerold is riding **Double Tough San** by Sak Em San out of OH April by Driftwood Ike. Double Tough has roping earnings of well over $100,000.*

Jaywood	*Casper*	*French Flash Hawk "Bozo"*
Kathryn Donavan	*(by Driftwood Ike)*	*(by Sun Frost out of Prissy*
Catty Peake	Mike Beers	*by Driftwood Ike)*
(by Speedy Peake)	Allan Bach	Kristi Peterson
Dave McGreggor	*Sam*	*Charlie (by Speedywood)*
Drift O Smoke	*(by Driftwood Ike)*	Joy Lynn Potter Alexander
(by Chipperwood)	Mike Beers	*Buck*
John Hoyt	Allen Bach	*(by Driftwood Ike)*
Black Bee	*Dance*	Cliff Whatley and
Larry Landsburg	*(by Driftwood Ike)*	Fran Whatley Snure
Speedy Bar	Mike Beers	*Porky*
Tuni Peake	Allan Bach	*(by Driftwood Ike)*
Gill	*Ike*	Mel Potter
(by Easy Keeper)	*(by Driftwood Ike)*	*Wilywood*
Jim Rodgriquez, Jr.	Clay O'Brian Cooper	*(by Orphan Drift)*
Easy Gran (Cadillac)	*Six Pac*	Tom Eliason
(by Easy Keeper)	*(by Speedywood)*	*Clyde*
Jim Peterson	Julio Moreno	*(unregistered by Driftwood Ike)*
Tee Woolman	*Redwood Man*	John Fincher
Jake Barnes	*(out of Swiftwood)*	*Lightning*
David Motes	Jake Kittle	*(unregistered by Driftwood)*
Don Gatz	John Hoyt	Glenn Motes
Jim Wheatly	*Booger Ike*	*Aliso Oso Rubia*
Joe Murray	*(by Driftwood Ike)*	*(by Aliso Quinton*
Snip	Barrie Beach Smith	*by Easy Keeper)*
(by Driftwood Ike)	*Dusty*	Jim Wheatley
George Richards	*(by Driftwood Ike)*	*Driftwood Mas*
Brad Smith	Barrie Beach Smith	*(by White Lightning Ike)*
Otoe Barwood	*Gilley*	Jim Wheatley
Walter Keefer	*(by Driftwood Ike)*	*Goldenwood Jaye*
Bar	Barrie Beach Smith	*(by Hesa Gill*
(out of a Speedywood mare)	*Flag (Paint)*	*by White Lightning Ike)*
Tom Ferguson	Sam McKinney	Wade Wheatley
Wildfire	*Shiney*	*Jake (unregistered)*
(by Speedywood)	*(by Driftwood Ike)*	Vern Castro
Reg Camarillo	Barrie Beach Smith	Frank Santos

cowboy went rodeoing with Riley Freeman of Baker, Oregon. Freeman was going to school, and rodeoing for, Cal Poly at San Luis Obispo, California. John had two horses with him, his seasoned rope horse Joe, and Chakaty, then a four-year-old started calf roping mount. She'd been roped on at home but had never been to a rodeo.

Riley Freeman rode Joe at an Idaho rodeo and got him over the rope and hurt, putting the horse out of competition. John was entered at two PRCA rodeos at Nampa, Idaho and Alturas, California. That meant that Chakaty got her baptisim under fire. The pair didn't win anything at Nampa since both horse and rider were frightened of the other ropers (Dean Oliver, Tuffy and Jimmy Cooper, and other world-famous hands) and the crowd. John roped at Nampa, loaded the mare in the trailer and drove all night to make the next afternoon's performance at Alturas, where they won first. Several ropers approached the young man about purchasing the mare but he turned them down, saying that he was not interested in selling. John was really "high" on her disposition and how she could run and stop. Chakaty was out of Lady Lux N., a Buck Nichols mare by My Texas Dandy. She was a great young mare with her whole future ahead of her.

Another Driftwood that Fincher regarded highly was Chilina, a daughter of Driftwood out of Hancock Belle. Don Nichols purchased Chilina from Katy Peake as a weanling, broke her to ride and roped on her until she was five when he sold her to Al Lauer.

John Fincher had left Chakaty with Don Nichols while he went on a two-year Mormon mission in Texas. Don was to breed the mare so he would have a colt when he returned. Don roped calves and heeled on her. Chakaty was a smaller type of mare and was jerked down a time or two in the rough, tough team tying event.

In 1960 Nichols called John in Texas and said that Al Lauer had offered to buy Chakaty for $5,000. John didn't want to sell the mare so he turned down the offer. Don later called and said that while Chakaty was a real good mare, he felt that Chilina would be better to team rope on because she was stouter. At Nichols' urging, Fincher put up Chakaty and Don anted $2,500 for Chilina. John owned two thirds and Don one third.

John Fincher returned from his mission in January of 1961 and enrolled at Brigham Young University in Provo, Utah. In late Febuary, some of the college cowboys began talking about the upcoming college rodeos that spring. John hadn't been on a horse for two years but decided to start roping again. He went to Arizona, purchased a pickup, picked up Chilina and hauled the mare to the NIRA Rocky Mountain Region rodeos.

The cowboy proved that he hadn't lost his touch with a manila and pigging string, winning the Intercollegiate Finals at Sacramento, California and making runner-up to the Championship on Chilina that year. He most probably would have won the year-end title for a second time except the hard-stopping mare broke a rope at the BYU rodeo earlier in the season.

Several of Fincher's rodeo teammates also rode the good mare that year, in both the calf and ribbon roping. Teammate Jerry McDonald of Tyron, New Mexico rode her at the BYU rodeo in the ribbon roping and gave the crowd a good laugh. The calf ducked back to the left and McDonald, a saddle bronc rider and bull dogger, didn't make the turn. He got lots of razzing from his teammates about not being able to ride a gentle saddle horse.

John and Don Nichols roped calves off of Chilina for several years. She had a great stop, could really

Fig. 8-3. Photo by Foxie; courtesy Mel Potter.

Two sons of Driftwood Ike in action at Yuma, Arizona in 1968. **Mel Potter** is heading on **Porky** and **Cliff Whatley** on **Buck**.

run and was double stout at the end of the rope. If a bull got out at a rodeo, Fincher could go rope the animal and pull it back into the arena with no fear that he or the horse would not be able to handle it.

After Chilina got older, Fincher purchased Nichols' interest and bred her to Speedy II, hoping for a filly. She delivered a stillborn foal so the cowboy traded her to Bob Ellsworth for a small bunch of Angus heifers.

John feels that if he had things to do over, he would have probably kept Chakaty instead of trading for Chilina in partnership with Don Nichols. Chilina was a great mare, although of a different type. She was much heavier and not as agile as Chakaty. But both mares could really run, stop and get back on the rope. Chakaty would have been a great dally heel horse today, but wasn't quite big enough to take the hard jerks that came with team tying. Like

Fig. 8-4. Photo courtesy Dr. Frank Santos.

Russ Santos heading on **Jake** at Salinas in 1958. The heeler is Frank Santos who owned the good bay gelding.

Fig. 8-5. Photo by Hubbel; courtesy Jim Wheatley.

Wade Wheatley *of Hughson, California turns one back on* **"Woody" (Goldenwood Jaye)** *at the 2001 National Finals Rodeo. The versatile palomino son of Hesa Gill Ike is equally adept at either heads or heels.*

Fig. 8-6. Photo by Hubbel; courtesy Jim Wheatley.

Jim Wheatley *of Hughson, California winning the 1995 Senior Steer Roping at the Ellensburg, Washington PRCA Rodeo. His mount—the capable* **"Ruby" (Aliso Oso Rubio** *by Aliso Quinton by Easy Keeper out of Lane J Bert).*

all the Driftwoods, both Chakaty and Chilina had lots of heart and worlds of try.

John Fincher has had Driftwood horses for 45 years (since 1956) and still has Driftwoods today. He has two daughters of Orphan Drift, a double granddaughter of Lucky Blanton, and two King-bred mares. He maintains that the Driftwoods could run and stop, had lots of cow sense and were stout at the end of the rope. They were great rope horses in the 1940s, '50s and '60s and are still outstanding rodeo horses today.

A father-and-son duo, Jim and Wade Wheatley of Hughson, California, have done right well roping off of Driftwoods for a number of years. Jim, a Top Fifteen PRCA team and single steer roper, has made several trips to the NFR, usually aboard a Driftwood-bred horse. His big taw in the tripping for a number of years was Ruby (Aliso Quinton by Easy Keeper and out of a Bert-bred mare). One of his better head horses was Bennie (Driftwood Mas), a bay mare by White Lightning Ike and out of Pudgy Can (Bold Britt II/Driftin Duchess by Easy Keeper). Son Wade has kept up the family tradition, both by being a world-class roper (PRCA Top Fifteen team roper and NFR contestant) and by riding a Driftwood. Capable of roping either heads or heels, Wade is mounted on Woody (Goldenwood Jaye) by Hesa Gill Ike and out of Berta Jay, combining a double dose of Driftwood with Frostys Tops.

A rodeo family which has ridden the Driftwoods with much success is the Beechs of Arizona and now Texas. The family—Sonny, Pat, and children Bret, Barrie and Beth—originally lived at Higley, Arizona. The close proximity to Roy Wales at Queen Creek and Driftwood Ike started the Driftwood chain while the three children were competing in High School rodeos. A roper himself, father Sonny knew the importance of being well mounted and he made sure that his children were.

Bret has been a world class team roper during his career. Although he didn't travel as heavily as many cowboys, preferring to stay close to home, he was always a threat when he backed his horse into the box and could rope horns or heels with equal ease. Barrie, the oldest daughter, ran barrels, poles, tied goats and team roped off of a dun gelding by Driftwood Ike to win multiple High School rodeo championships. She is now married to former World Champion Team Roper Brad Smith and raises Driftwood horses at Stephenville, Texas. Beth, the youngest daughter, also competed and won riding Driftwood Ike progeny. She is married to seven times World Champion Team Roper Clay O'Brien Cooper and lives in Granbury, Texas. Clay has ridden the Driftwoods consistently during his long tenure in the arena and feels that "they're the best."

Once a horseman has ridden the Driftwoods it seems to become a family habit. As children grow up, begin to compete and need to be mounted, the majority of ropers who have been on a Driftwood make the same choice in horseflesh for their offspring.

Fig. 8-7. Photo by Les Walsh; courtesy Dr. Frank Santos.

Frank Santos *steps off of* **Jake,** *a Driftwood-bred gelding, at the 1954 Clements, California Stampede. Santos was .16 seconds on this run. In Frank's words, "Jake was awesome."*

Buck Nichols started the chain, putting his sons Don and Hugh on Driftwood-bred contest mounts...and then repeated the process with his grandchildren. Asbury Schell was quick to see that his son Eddie was aboard Cowboy, Maestro and Cherokee Jake, all by Driftwood. Roy Wales kept his children mounted on sons and daughters, grandsons and granddaughters of Driftwood, such as Driftwood Ike. Will Gill, Jr., after owning Easy Keeper, purchased White Lightning Ike who sired good roping geldings for son David. Henry Kibler of Arizona started out as a roper, got the Driftwood bug and passed it on to his children, who still ride Driftwoods.

The Driftwood tradition continues with the children of Tom Eliason. They compete and win on sons and daughters of Wilywood, the stallion on which Tom won so well. The Arnold and Pixley families are also making a habit of mounting their sons on both ranch stallions and the sons and daughters of those stallions. In northern Arizona Vic Howell, cow boss of the sprawling CO Bar, uses Driftwood-bred geldings as ranch mounts, as well as putting his children a-horseback at youth rodeos.

The list goes on and on, with many young cowboys and cowgirls growing up aboard a Driftwood-bred horse.

9

THE LINE CONTINUES

"There was something in that Driftwood breeding. His colts were quiet, athletic and would do whatever you wanted them to do. There will never be another stallion like him."
—George Seals

Double Drift
Orphan Drift
Shady
Six Pac
White Lightning Ike
War Drift
War Concho
Lone Drifter
Speedy Man
Hug Me Chick
Redwood Man
Cibecue Roan
Easy Gran (Cadillac)
Jimmywood
Wilywood
Blue Light Ike
Drift Chip
Lindsay Peake
Otto Barwood
Goldenwood Jaye (Woody)
French Hash Hawk (Bozo)
Isle Drift

The story didn't end with Driftwood's sons and daughters. They had inherited his genes and ability to pass on the performance potential to their get. That trait, more than anything else, has made the Driftwoods sought after for half a century. As Roy Wales was so fond of saying, "If a horse has even one drop of Driftwood blood, he's got to be good."

GRANDSONS:

Double Drift This black stallion, bred by Tex Oliver and foaled in 1956, carried a double dose of Driftwood, being by Gray Chip (Driftwood) and out of Rosewood (Driftwood). Freeland Thorson first saw the stallion at Rancho Jabali when he was a two-year-old. Thorson was with Bill Gibford, head of the Horse Department at Cal Poly. At Katy Peake's urging, Gibford, who had ridden several Driftwood foals, selected a two-year-old colt to train. That colt was Double Drift. (Speedy Peake was in the other corral.) Freeland thought Gibford chose the colt because of the double infusion of Driftwood blood.

"Somehow, we got him in that old, green quarter-top trailer that the college owned," remembered Thorson. "The colt was frightened to be by himself and we had every kind of rope imaginable on him to keep him from climbing out. That's when we stuck the nickname "Boogie" on him. Well, we took him back to San Luis Obispo where Gibford started his education. Bill was really a top hand and believed in the Driftwoods."

When Katy Peake decided to leave the horse business, Bill Gibford still had the young stallion in training. Although she had been approached by some Texans who wanted to buy him, Katy preferred that Double Drift stay in California. She contacted Doc Whitman, Thorson's father-in-law, about purchasing the stallion. Whitman did so—and sent him to Monty Roberts for cutting training. The horse was later moved to the Whitman ranch at Lodi.

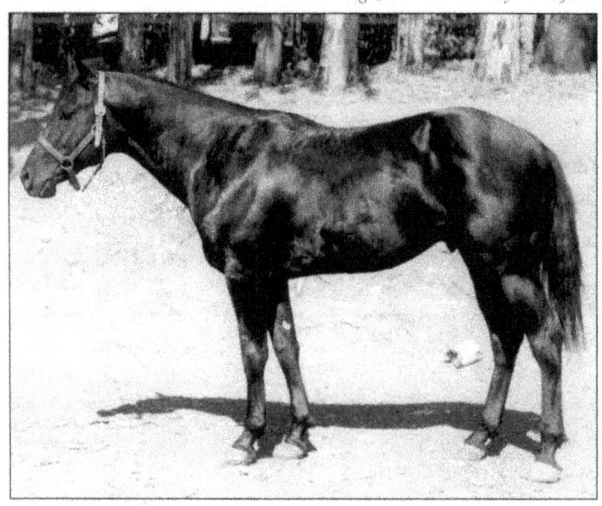

Fig. 9-1. Photo courtesy Peake files.

Double Drift 85080 *This black stallion, by Gray Chip by Driftwood and out of Rose Wood by Driftwood, was bred by Tex Oliver. He was a versatile performer, earning AQHA points in six events as well as money in NCHA-approved cuttings. He was also an outstanding team roping mount—and a winner at both the big jackpot ropings and PRCA events. As a sire, he passed on his athletic ability and willing disposition.*

Double Drift had a stellar show career, winning AQHA points in Halter, Heading, Heeling, Barrel Racing, Working Cow Horse, Cutting, and Western Pleasure. He was awarded his Open Performance ROM in 1967. He was also a winner at National Cutting Horse Association contests. Since Whitman liked to team rope, Boogie spent a lot of time in the roping arena and distinguished himself as a top heading horse. In 1962, at the Oakdale, California 10-head team roping, Bud Corwin (on Double Drift) and Doc Whitman were 6.3 and 7.1 on their first two steers over the long score. One story Whitman enjoyed telling was the day he saddled Double Drift to rope a few practice steers. He rode into the arena and kicked Boogie into the "team roper warm-up lope." What he didn't notice was that his daughter Diana had left the barrels set up from the previous day's practice session. When Double Drift got close to the barrel, he went around it. Doc Whitman kept going straight ahead—into the dirt.

In 1967, since he was selling his ranch at Lodi, California, Whitman gave the stallion to his daughter, Diana Thorson. His new home became the Bar 7X Ranch at Weiser, Idaho.

Double Drift sired 171 foals, including AQHA Champion Cowpoke Pete; AQHA Champion Hondo Drift; AQHA Superior Youth Mare and Superior Western Pleasure Horse Driftwood Micky, who accumulated the huge total of 608 AQHA points in both Senior and Youth competition; Double Shady; Double Shad; Drift's Diplomat, shown in Snaffle Bit Futurities and Cutting, a top barrel horse and rope horse that is retired to raise Will and Cheryl Hall's three children (the Thorson's grandchildren). Other notable foals included Nickie Pine, dam of Isle Drift; Drift's Darlin'; Drift's Dariene; Dynaflow Drift, the barrel and team roping stallion owned by the Halls; and Drift's Chip, a rope horse, NFR pickup horse and outstanding sire, first for Jim West and then for Perry Johns.

In looking back over the years she has been riding Driftwood horses, Cheryl Hall has lots of good memories and a few heartwrenching ones. "I've always said that the Double Drift offspring have a heart of gold and Drift's Darlin' (DeeDee) was not exception. At 18 she was still going strong, leading the barrel racing in her Inter-Collegiate Rodeo Region when she was injured and had to be retired. I was at a college rodeo and second in the long go with the Finals that night. DeeDee had never been lame a day in her life, but that night she was limping. Assuming a stone bruise because of the rocky ground, I gave her some Bute and ran in the short go, winning it and the average. Later we learned that she had torn a hole the size of my little finger in her suspensory tendon. Since I always kept her legs wrapped on the road, I didn't detect any swelling or heat until we got home. It still brings tears to my eyes when I remember the heart that mare exhibited during her final barrel race."

"Dynaflow Drift (aka "BF"), a full brother to Drift's Darlin', is the youngest of the three Double Drift sons left. He stands just shy of 16 hands and weighs 1,400. Like his father, he is an all-around horse with a wonderful disposition, great conformation with big, strong bones and feet which he passes on to his foals. His breeding is somewhat different than the majority of the Driftwoods, being out of King Bunny King by Country Boy Lauro, a King Ranch-bred stallion, which makes him a great cross on Driftwood Ike-bred mares."

Thorson's last comment was, "I have been told that Double Drift was a 'catch colt.' If so, we'd like to have a truck load just like him."

Orphan Drift 240825 A dun stallion, foaled in 1962 by Driftwood Ike out of Orphan Annie C, has done much to carry on the Driftwood name. Bred by Roy Wales and purchased by Stanley Johnson, he sired numerous good geldings and broodmares as well as show winners. Lady Drift, a 1965 roan mare out of King's Ellanita, earned her Performance ROM in 1970, placed 6th in the 1975 Open World Champion Senior Reining, and 5th in the 1975 Open World Champion Working Cow Horse. Cinderwood Miss, a 1969 daughter of Poco Judy Sue, achieved her Open ROM in 1974. Flintwood, a 1971 gelding out of Poco Fernwood, was 3rd High Point Youth Breakaway Roping, 6th World Champion Youth Calf Roping and 4th Youth World Champion Breakaway Roping in 1979. In 1980, the buckskin gelding earned his Open ROM and was 10th in the World Champion Team Roping. Drift's Super Star, a 1975 gelding, won his Youth ROM in 1981. The 1975 gelding Geowood won a Youth ROM in 1981, and Triple Drift, a 1977 gelding, achieved the same honor in 1984 as well as being a money-winning team roping mount.

Perhaps his best-known son is Tom Eliason's Wileywood out of Oui Oui (Poco Speedy/ Woodfern by Driftwood).

Shady Purchased in 1955 by Dale Smith of Chandler, Arizona, this unregistered black son of Cherokee Jake out of an Ellsworth mare was an outstanding roping horse. In twelve years on the pro rodeo circuit, Shady carried Smith to a pair of World

Fig. 9-2. Photo by Stanley Johnson.

Orphan Drift, by Driftwood Ike and out of Orphan Annie C. was owned by Stanley Johnson. He did much to carry on the Driftwood line by siring outstanding show and performance horses, including Wilywood.

Fig. 9-3. Photo by DeVere.

Shady, by Cherokee Jake, was one of rodeo's greatest rope horses during the years that Dale Smith hauled him. Smith won two PRCA World Championship titles on the stout black gelding.

AQHA Get Performance Summary
ORPHAN DRIFT

Total Registered Foals	249
Performers	27
Crops	22
Starters	2
Total Starts	12
Halter Point Earners	4
Halter Points	6
Working Point Earners	22
Working Points	234
Registers of Merit	6
Performers	23

Champion Team Roping titles. The Arizona cowboy headed, heeled (in both dally and tie-down versions of the event), roped calves and tripped steers off the good gelding. Homer Pedigrew won a 3rd in the Calf Roping off of him at the big Los Angeles rodeo one year.

According to Smith, "Shady would rather pull than eat and didn't know just how stout he was on the end of a rope." One year at the Chandler, Arizona team tying, Smith was heeling for John Clem. John headed the steer and "went left." Smith roped one hind foot, spun around and turned Shady loose. The black gelding went to the end of the nylon—and broke it! The rope flew back, hit Smith on the right forearm and fractured it.

With his pair of Driftwoods (Shady and Poker Chip Peake) in the trailer, Dale Smith was a threat to other contestants in both calf and team roping every time he pulled up to a rodeo.

Emm Dee 244374   This brown 1962 son of Speedywood and out of Mat (known as Cinch when Dick Johnson was roping calves on her) by Texas Tom Boy did what the Driftwoods do best—rope. He was originally owned by Dick Johnson, who started rodeoing on the 14-hand, 1070-pound stallion when he was a four-year-old. Johnson also bred him, putting approximately fifty foals on the ground by the time the horse was seven. Emm Dee was then purchased by the capable Mike Quick of Hemet, California, who had him gelded and continued to rope on the horse. In 1973, when Emm Dee was eleven, Pete Zanette, now of Weatherford, Texas but living in southern California at the time, bought him. Pete got his money back many times over. A believer in the old rodeo adage of "finding a horse that fit him and then staying on it," Zanette hauled the consistent performer until he was 27. During Emm Dee's long career he carried ropers to win or place at every big PRCA rodeo on the Pacific Coast from Pendleton, Oregon all the way south to the Mexican border.

"He was an outstanding horse," recalled the California hand. "He scored like a dream, could catch cattle, stop and really work a rope. And he'd been hauled so much that he could sleep in the trailer. At one time or another, just about all the tough ropers rode him and won. He had lots of heart, too. One year at the Vallejo, California rodeo

Fig. 9-4. Photo by Hyder; courtesy Pete Zanette.

Emm Dee, *a 1962 son of Speedywood, comes to a stop with owner* **Pete Zanette** *stepping off at the 1977 Springville, California rodeo. Zanette drew this same calf three rodeos in succession, roped him off of Emm Dee, and won over $2,600. The capable brown gelding was rodeoed on for 23 years.*

another horse crowded him and he fell on top of a track harrow, tearing a big hole in his side. The horse was layed up up the University of California Veterinary Hospital at Davis for four months, but came back ready and willing to work. Emm Dee was the kind that you always remember."

Six Pac (registered as Kool Pack 806555)   This brown 1968 son of Speedywood was out of Tic Nac by Tick Tack Darter. He was bred by Dale Smith and went to rodeo's big leagues as a head

Fig. 9-5. Photo by James Fain, courtesy Julio Moreno.

Six Pac *(right) turns a steer for* **Julio Moreno** *at the 1976 NFR at Oklahoma City. Six Pac was one of the top head horses of his day. Heeler is Dennie Watkins.*

horse under Julio Moreno, now of Marysville, California. During his years on the road, there was no major team roping or big PRCA rodeo where money wasn't won on the capable gelding at one time or another. Like most of the Driftwoods, Six Pac scored well, could catch cattle in a hurry and was capable of smoothly turning a big steer at the end of a rope in money-winning time. For several years the good gelding and Moreno were a regular team at the National Finals Rodeo when it was held in Oklahoma City, Oklahoma.

Top performance geldings like Six Pac are the reason why the Drifwoods have their well-deserved reputation.

White Lightning Ike 1650151 This 1980 buckskin stallion sired by Driftwood Ike out of Katy Was A Lady by K4 Hickory Skip, 1964 AQHA Champion. He was bred by the K4 Ranch at Prescott, Arizona and purchased by Will Gill & Sons of Madera, California as an eight-year-old. He has sired numerous ranch, rodeo and barrel horses. In addition to good minds and athletic ability, White Lightning Ike sires lots of color, duns and buckskins, which is always in demand. Other than a handful held back for broodmares or ranch geldings, the majority of his foals are spoken for as weanlings and sold by the time they are yearlings.

War Drift 434663 This brown stallion, foaled in 1966, was by War Chief by Joe Hancock and out of Brown Beulah by Driftwood. He was bred by W. G. "Tex" Oliver, who sold him to Flint Fleming and Jean Morford of Miles City, Montana. From there he went to the Haythorn Land and Cattle Company at Arthur, Nebraska, where he was used as a stallion for ten years. A son, WD Five, has been used by the Haythorn Ranch extensively to sire ranch and rodeo horses. War Drift was finally sold to Markwood Enterprises of Chandler, Arizona.

War Drift and his full brother War Concho have proved the effectiveness of the Joe Hancock/Driftwood cross in the production of usin' horses.

Fig. 9-6. Photo courtesy Will Gill & Sons.

White Lightning Ike, *a buckskin son of Driftwood Ike out of Katy Was A Lady, has sired numerous top performance horses. Bred by the K4 Ranch, "Ike" has been owned for most of his life by Will Gill and Sons of Madera, California.*

Fig. 9-7. Photo courtesy Henry Kibler.

War Drift, *by War Chief by Joe Hancock and out of Brown Beulah by Driftwood. This brown stallion made substantial contributions to the working horse world.*

Fig. 9-8. Photo courtesy Menefie Hill Ranch.

War Concho, *by War Chief out of Brown Beulah by Driftwood, was bred by Tex Oliver. He was used extensively as a stallion.*

War Concho A full brother to War Drift, also bred by W. G. "Tex" Oliver, was by War Chief by Joe Hancock and out of Brown Beulah by Driftwood. War Concho was used by Charlie Judd, and later Henry Kibler, to carry the tradition of producing good ranch and rodeo mounts.

Lone Drifter This buckskin son of Driftwood Ike and out of mare by Yendis was from Ike's last colt crop. He was purchased by Mel Potter from the K4 Ranch at Prescott as a three-year-old. Lone Drifter had been injured as a foal and was not sound enough to ride, so was immediately turned out with a friend's mares. As a sire, the buckskin stallion has more than proved his worth. For several years, Potter allowed friends to utilize the horse, taking a couple of colts or fillies annually as payment. Eventually, he accumulated a select band of Driftwood and Driftwood-influenced mares, brought Drifter home, and was in business.

Fig. 9-9. Photo by Jim Morris.

Lone Drifter, *by Driftwood Ike, is the senior sire at the Potter Ranch at Marana, Arizona. This buckskin stallion has sired many good rodeo and show mounts.*

Over the years, Lone Drifter has sired top usin' horses, contest and show mounts, and producing daughters which have carried on the tradition of the Driftwoods.

Speedy Man A 1963 bay stallion, by Speedywood by Driftwood and out of Maybelline by Red Man and out of Brown Beulah by Driftwood, was bred by Maynerd Gayler of Aria Vista, Arizona. Gayler, a friend of Dale Smith's, used Speedywood for several years as well as owning Maybelline. Speedy Man was utilized as a ranch sire until the Gaylers had too many of his daughters. He was then sold to a ranch across the border in Mexico, remaining there until 1978.

That year Harper McFarland, manager of the San Christobol Ranch at Lamy, New Mexico acquired him and raised more ranch horses. In 1984, when he was 21, Speedy Man was purchased by Henry Kibler of Chandler, Arizona. The bay stallion produced six more colt crops, dying at 29 in the spring of 1992.

Henry Kibler remembered the first time that he saw Speedy Man. "He was not a big, heavy horse in any way, weighing about 1075 pounds, not real deep in the chest, and had a nice, big hip. He had a nice slim neck and sloping shoulder. His head was nice, although certainly not a 'halter horse' head by any means."

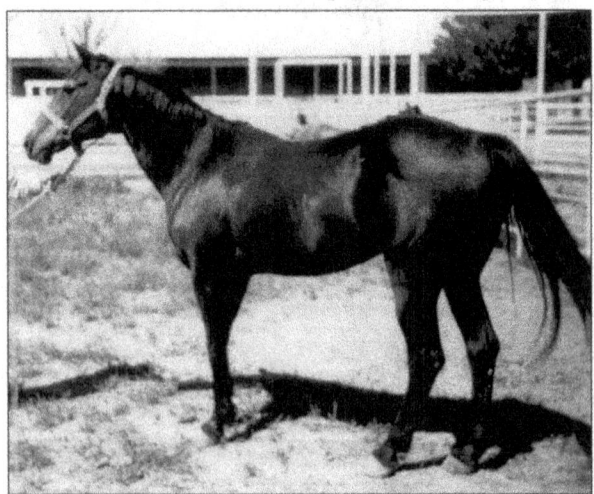

Fig. 9-10. Photo courtesy Henry Kibler.

Speedy Man, *by Speedywood out of Maybelline (Red Man/Brown Beulah). Bred by Maynerd Gayler, he was used as a stallion by his owner, a Mexican ranch, the San Christobol Ranch, and Henry Kibler.*

Kibler continued, "A friend, Randy Johnson, had located two Cibecue Roan daughters at the San Christobol Ranch and called me about them. We made arrangements to meet and set up an appointment. When we arrived, McFarland was not there yet and Randy McCorkel showed us around, pointing out some young geldings he was riding. I especially liked two brown horses, about 15 hands plus, fairly stout and who really moved well. Randy said that they were as good as any he had ever ridden. He then showed us some other horses and a few colts. Naturally, I asked how they were bred. He said that they were by the same stud as the two geldings—a horse called Speedy Man—a Driftwood horse."

"About that time Harper showed up, and after some discussion, I bought the two mares. Then I asked about the stallion. Harper told me about him and then said, 'Well, heck, let's go look at him. He's out with some mares and if you don't mind a ride in the pickup we can find him.'"

"We pulled into a three-section pasture, found the mares, and I asked Harp where the stud was. He pointed out a medium sized brown horse and said 'That's him.' I couldn't believe that he was the sire of those geldings back at the ranch. Of course, he'd been out with 15 mares that season, needed a good rest and grub, but he wasn't much to look at that day."

Kibler went on, "I told Harper that I had a few Driftwood mares and that I'd be interested in breeding to the horse. We worked out a deal for the next year."

"That fall," continued Henry, "Harper called and asked if I was interested in buying Speedy Man. The horse was doing pretty well but Harper was afraid that he wouldn't winter well in the high country and would do better in a milder climate. I jumped at the chance and a couple of weeks later brought him to his last home here in Arizona."

Among the good sons of Speedy Man were Speedy Roan Man, Drifter Joe Wood, Setemup Speedy, Red Roan Speedy and Triple Drift.

Hug Me Chick   This versatile individual was sired by the sensational Triple Chick (Three Bars/Chicado V) and out of Hug Me Tight (Driftwood/Nugget Hug). He was a champion Stock Horse in California when shown by Jim and Cherri Alderson, as well as earning AQHA points in Western Pleasure and Working Cowhorse. Hug Me Chick was also a money-winning cutting horse. Often on his way to a cutting he would be unloaded at a roping to carry his owner to either the horns or heels. When the roping was over, he'd go back in the trailer and go on to the cutting, where he frequently placed.

Hug Me Chick foals have accumulated both working and halter points at AQHA shows, have earned a

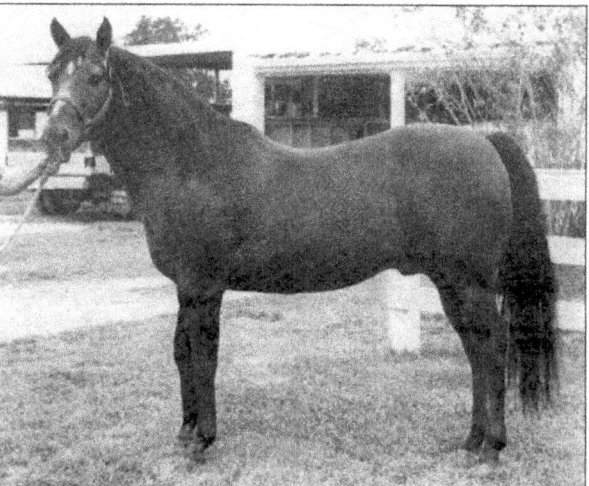

Fig. 9-11. Photo reprinted from Quarter Horse News, October 25, 1983.

Hug Me Chick, *a Driftwood grandson who could do it all—stock horse, cut, rope and pleasure—and passed it on. Sired by Triple Chick by Three Bars, he was out of Hug Me Tight by Driftwood.*

ROM, and have successfully run on the race track. Daughter Hug Me Chiquita has 61 working points, cutting points as well as a ROM and a Superior in the event, as well as placing 7th at the 1975 World Championship Senior Cutting.

Another daughter, Hug Me Chickadee, was an NCHA Futurity competitor as well as earning AQHA cutting points. She was the dam of Lena Chick, 1982 Super Stakes finalist.

Hug Me Chick was purchased from the Aldersons in 1975 by the Village Creek Ranch at Burleson, Texas.

Red Wood Man P-75573   This versatile stallion, an Open Performance ROM winner, bred and owned by Jake Kittle, was an impressive performer. In 1961 he was the Grand Champion Working Horse in Arizona in addition to being Champion Reining and Working Cowhorse and Reserve Champion Calf Roping Horse. He was by Red Man and out of Swiftwood by Driftwood, a full sister to the great Jernigan Peake who was the Champion Open Stock Horse in the California Reined Cow Horse Association. He combined the proven performance ability of Joe Hancock and Driftwood. As a sire, he passed on his impressive ability.

Cibecue Roan P-99541   This roan stallion was by Red Man and out of Miss Linwood by Driftwood and bred by Jake Kittle during the time he had Red Man leased from Kenneth Gunther. Under Grady

Fig. 9-12.

Red Wood Man, by Red Man out of Swiftwood, was not only an outstanding performer himself but passed the ability on to his get.

Fig. 9-13.

Cibecue Roan, by Red Man and out of Miss Linwood by Driftwood. He was bred and owned by Jake Kittle and exhibited by **Grady Stewart.**

Stewart, the Hancock-Driftwood stallion demonstrated his ability as a performer by earning a Performance ROM and Open High Point Steer Roping Stallion in AQHA competition. Most of his competition was confined to the Inter-Mountain area, but when he was shown, he dominated his classes. As a sire he produced numerous talented geldings and broodmares who contributed to the performance horse legacy.

Easy Gran (Cadillac) An aptly nicknamed bay gelding, by Easy Keeper and out of a Red Mud daughter, was foaled in 1974 (the year Easy Keeper died) and bred by Will Gill & Sons of Madera, California. He developed into an outstanding head horse, carrying Jim Peterson and David Gill to the 1985 NFR. Jake Barnes also rode him there to cinch the first of his seven PRCA World Champion Team Roper titles. He has also been ridden by World Champions Tee Woolman and David Motes in addition to top fifteen ropers Don Gatz and Jim Wheatly. His last owner, Joe Murry of Oakdale, California, tripped steers on the dependable gelding up into his twenties. Those ropers who competed on him all agreed that "he was a Cadillac."

Jimmywood P-207548 This 1962 black son of the fast Super Charger and out of the sensational Henny Penny Peake fulfilled the predictions of his performance potential. Bred by Al and Norma Lauer, he combined the Driftwood heritage with the speed of Super Charge, plus a dash of King through his granddam O'Quinn's Midget. Under the capable Ronnie Richards, Jimmywood developed into an outstanding stock horse, who among his other accomplishments took the 1966 Santa Barbara Nationals championship. He had the style, class, and athletic ability that made him a favorite with his riders and the judges.

Fig. 9-14. Photo by Dan Hubble; courtesy Joe Murray.

Joe Murray tripping steers on **Easy Gran (Cadillac)** when the horse was 23 years of age. Cadillac was by Easy Keeper by Driftwood.

Fig. 9-15. Photo courtesy Peake files.

Jimmywood, a son of Henny Penny Peake by Super Charge, went on to be an outstanding stock horse.

Fig. 9-16. Photo courtesy Eliason family.

Wileywood, a double-bred Driftwood by Orphan Drift out of Oui Oui, showing his ability as a rope horse <u>without a bridle</u> for **Robert Eliason.** Owned by Eliason Quarter Horses. Wilywood has sired AQHA, ROM and World Show qualifiers; barrel futurity and derby money winners; PRCA, amateur, high school and 4-H rodeo winners; NRHA, NBHA, USCRA and USTRC money winners; FQHR National Show Champions and winners.

Great Grandget:

Wilywood A 1977 son of Orphan Drift by Driftwood Ike and out of Oui Oui by Poco Speedy and out of Woodfern and is owned by Tom Eliason of Gregory, South Dakota. Eliason was first introduced to the Driftwoods when he was in college. His roommate, John Fincher, won the Intercollegiate Calf Roping title in 1964 riding a son of Driftwood. When the opportunity arose to purchase the six-year-old dun stallion Eliason traded another horse and $200 for him. As a contest mount for the calf-roping Eliason, Wilywood has carried him into the money at numerous contests. In AQHA competition, Wilywood has collected points in Reining, Calf Roping, Working Cow Horse, and Team Roping, and wasn't retired until he was nineteen. His foals have been winners in the same events, as well as pole bending, barrel racing and team penning, and have performed well in the rodeo arena.

Eliason remembers keeping the horse in a small, rented pasture behind his house at first. "It was either that or buy a couple of steers to keep the grass down." Wilywood not only kept the grass mowed but brought in some much-needed money from stud fees. A $150 stud fee was a lot more than I'd have gotten for a steer, and we bred more than one mare."

Tom Eliason continued, "Wilywood has done it all himself. We weren't able to put a lot of promotion money behind him and he sired good foals out of just about any kind of mare. We started with a $150 stud fee and the mares just kept coming. Raising it didn't make any difference in Wilywood's popularity as a stallion. His 2001 stud fee was $2,500 and he was booked full."

"The most interesting thing we find with the people who are standing sons of Wilywood is that they all think they have his best son." Tom says. "They can all give you a long list of reasons why they think they are right."

Twelve of those sons are Wily Wooddrifter, Wily White Sock, Lucky Goodwood, Bull T Wood, Rabbit Choker Drifter, Wily Rox, Orphan Bear, Poco Woody Pine, Dandy Driftwood, Wily Tuko, Dakota Drift, and DVA Maxi Driftwood. All have carried on their sire's record as outstanding performance horses and as sires.

The Eliasons maintain a select band of mares and raise about 25 foals a year. Most go to ranchers and ropers, the people who use and ride horses. In August of 2000, the family operation held their first production sale. The top selling animal was a 1994 bay stallion, Wily Rox by Wilywood and out of Bay Lady Rox, who went for $24,900. Another Wilywood foal, a 1996 red dun filly consigned by Clair Jones, brought $15,000. The Eliasons hold their annual production sale in August.

Tom added, "We have told you about Wilywood and what he has done for us economically and as a performer and sire, but let me tell you what he means to our family. He is the kindest horse that ever was. He was patient with our kids when he taught them to rope and never cheated or messed them up. When the kids brought their friends out, they rode Wilywood. He was the safest horse on the place. Now we put a saddle on him and let the grandkids ride him around in back of the barn. How do you express your appreciation to a horse like that?"

Blue Light Ike This good looking, gray son of White Lighting Ike has done it all. Owned by Joe and Cathy Murry of Oakdale, California, who purchased him from Will Gill and Sons as a baby, Blue Light Ike has earned ROMs in Heading and Heeling, points in Team Penning and qualified for the AQHA World Show in two events. Additionally, he has been rodeoed on, winning at the PRCA level in addition to carrying his riders to the pay window at USTRA, NHSRA, NIRA and junior rodeo events. His athletic ability is matched by his disposition, making it possible for children to compete on the horse at youth contests.

As a stallion, Blue Light Ike is siring the kind of horses that are sought after by ropers and barrel racers up and down the Pacific Coast. A son, War Black Ike, owned by Gilbert Reynolds of Oakdale, California, is following in his sire's hoofprints, both as a rope horse and as a stallion. Blue Light Ike is carrying on the tradition of his sire, grandsire and great grandsire.

Drifts Chip A black stallion bred by Freeland Thorson of Nampa, Idaho, Drifts Chip was sired by Double Drift out of Diamond Isle by Diamond Chip by Silver King and out of a daughter of Dusty Hancock. This cross combined a double dose of Driftwood with the proven performance bloodlines of the King Ranch and Joe Hancock.

Fig. 9-17. Photo by Bob Wagoner-Sandy Skaar; courtesy Joe Murray.

Blue Light Ike, a son of White Lightning Ike by Driftwood Ike by Driftwood, carrying **Troy Murray** to the pay window at the Oakdale, California rodeo. Blue Light Ike has AQHA points in Heading, Heeling and Team Penning, and qualified for the World Show in two events. Additionally, he has been a winner at PRCA, USTRA, NHRA, NIRA and junior rodeos in all timed events.

Drifts Chip, a black 1979 son of Double Drift/Diamond Isle, has been a great ranch horse, rope horse and pickup horse, in addition to being a standout sire for Jim West of Ione, Oregon. In 1997 Drifts Chip was sold to Perry Johns of Fort Worth, Texas, who stands him at the 6666 ranch in Guthrie, Texas.

Freeland had an agreement with Witch Holeman of Nevada to break a number of his foals each year and Drifts Chip was one of them. Thorson remembers, "The night I hauled Drifts Chip down to Witch, he was a rank long yearling. Holeman had to rope him and jerk him down to get a halter on. That was the last time I saw Drifts Chip. Holeman called me some time later and said he had a buyer for the black colt. I said, "sell him!"

The purchaser was Jim West, who ranched at Ione, Oregon. Under his handling, Drifts Chip became an all-around ranch horse who worked for his feed all week, and a team roping and hazing horse on holiday. He also developed into a top pickup horse, seeing action at the National Finals Rodeo in Las Vegas. West described the black stud as "Pretty enough to lead a parade, tough enough to pick up bucking horses at the NFR, fast enough to haze for bulldoggers and still sound at 21 after years of hard use." Drifts Chip sired many good ranch horses in addition to roping and dogging mounts, barrel racers and show winners. His daughters are sought after as broodmares.

In 1997 West sold the stallion to Perry Johns of Fort Worth, Texas. Johns made arrangements to stand Drifts Chip at the well-known Burnett Ranch (the 6666) where the Texas horsemen would have access to him. During Drifts Chip's time in Texas, ranchers, rodeo hands, and breeders have taken mares to him. Some of these individuals are looking for performance geldings, others are interested in future broodmares or stallion prospects. As a sire, Drifts Chip has not disappointed them.

Lindsay Peake 2643854 Bred by Russell Keeley and owned by Jim Morris of Exeter, California, this brown stallion was sired by Son of Bar Girl by Son O Sugar and out of Sandy Peake by the versatile Speedy Peake by Driftwood. His breeding is a combination of the greatest performance horse bloodlines in the Quarter Horse breed. Shown in the 1990 Snaffle Bit Futurity, the athletic stallion performed creditably. He went on to win reinings, both in Hackamore and Bridle, as well as Working Cow Horse. He was also a ranch and rope horse before being retired to stud duty. As a sire he has put numerous good cow horses and brood mares on the ground. Several of his colts have been ridden by top charros in Mexico as well as the United States, and demonstrated their athletic ability in those demanding contests.

Fig. 9-19. Photo courtesy Jim Morris.

Lindsay Peake, *a brown stallion by Son of Bar Girl by Son of Sugar and out of Sandy Peake by Speedy Peake, combines some of the greatest performance blood in the Quarter Horse. He has foals in 17 states and they are making a name as doin' horses.*

In addition to heading Jim Morris' select band of Driftwood mares at Exeter, California, Lindsay Peake has been utilized one season by Gene Moench in Indiana, another year by John Balkenbush in Montana, and later by the Babbitt Ranches in Arizona. Lindsay Peake has sons and daughters in 17 states, a number of which are just beginning to make their marks under saddle.

Fig. 9-20. Photo courtesy Phil Livingston.

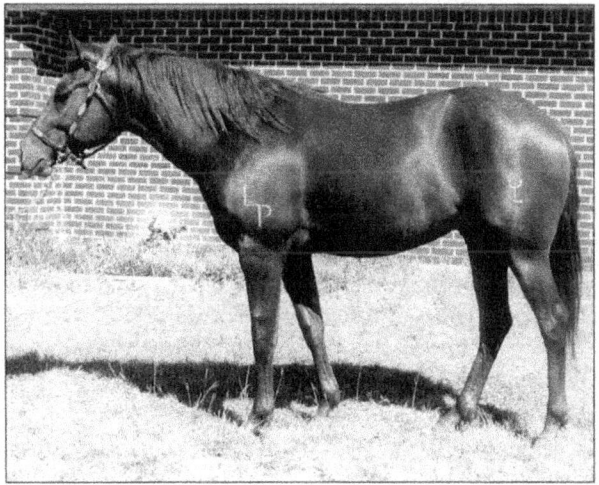

Rojo Colorado Peake, *a 2-year-old son of Lindsay Peake out of Lotta Driftwood (Speedyman/Senora Peake). Like all the Driftwoods he is built to perform.*

In each case, he has carried on the Driftwood tradition of siring horses that can get the job done under saddle and later produce others which will do the same.

Cowboy Drift This double great grandson, by Orphan Drift by Driftwood Ike and out of Poco Judy Sue out of Judy Sue by Driftwood, has been the senior stallion at the Babbitt Ranches at Williams, Arizona for a number of years. Crossed on a band of Driftwood-Hancock mares, Cowboy Drift has produced the type of horses that can rodeo and ranch. The Babbitt horses have the strong feet and legs, functional conformation and toughness to carry a man all day long in rough country and come back for more the next day. Realizing that the broodmare base is of prime importance, the Babbitt Ranches have made sure that every individual has one or more brothers in the remuda that have distinguished themselves under saddle. Ranch management knows what the foals are capable of before they are offered for sale.

The Cowboy Drifts are popular with ranchers, ropers, rodeo hands and rolks that just like to ride a good horse. His daughters are outstanding broodmares.

Otoe Barwood A Driftwood who made quite an impression on a Yankee cowboy back in the middle 1980s was Otoe Barwood. Walter Keefer of Lafayette, New Jersey was a regular competitor at East Coast rodeos as a roper and bulldogger. He trained his own horses and was always looking for a likely prospect. He'd heard of the Driftwoods through the rodeo grapevine but hadn't had any experience with them.

Some friends showed up with a classy looking blaze-faced, stocking-legged sorrel gelding. While the horse had been bred in Texas, somehow he'd made the trip East. Keefer liked what he saw and promptly made an offer—which was turned down. A year later, after discoving that the colt was always two steps ahead of them and had gotten his bluff in, the owners offered him to Walter. The cowboy took him off their hands—at a killer price. Looking at the registration papers Keefer discovered that his new purchase (Otoe Barwood) was by a grandson of Otoe and out of Lady Barwood by Wooden Nugget by Driftwood.

"Barwood had chilled out his previous owners by going in the air. Then, they made the mistake of

Fig. 9-21. Photo courtesy Walter Keefer.

Otoe Barwood, *an '83 son of Baldy Otoe out of Lady Barwood by Wooden Nugget, and* ***Walter Keefer*** *of Lafayette, New Jersey. After a rough beginning this blaze-faced sorrel gelding made believers out of East Coast cowboys about the Driftwoods.*

tyin' the spooky so-and-so up to a gate. He set back, jerked the gate off the hinges and took off down the road. When they finally got him cornered he had not only torn up the gate but the new saddle he was wearing. They called and said, "Come get him!" The first thing I did was put on a running W on and lay him down," remembers Keefer. "A couple of hours thinking things over and he came off the ground with a totally different look in his eye and a willingness to please that surprised me. We only had one reoccurrence of his wanting to rate in the box. And you could feel that it was only a half-hearted attempt. We went back to the ground for a reminder and never had any more trouble."

"Not only did the horse have ability under saddle, but he was a looker," commented Keefer. "Just riding him into the arena caused heads to turn, even among the show horse crowd. One halter horse man was comparing Barwood with his show winner and discovered that he had a bigger and better hind leg than his horse."

"I don't think that I ever had a horse that took to cattle like Barwood did," laughed the Jersey hand. "He had it all—was an athlete, could run, stop and really wanted to please. I roped calves, bulldogged

Fig. 9-22.

Goldenwood Jaye (Woody), *with owner/trainer* **Wade Wheatly** *in action at a PRCA rodeo.*

Fig. 9-23. Photo by Jennings.

French Flash Hawk (Bozo), *a Driftwood-bred bay gelding who has more than distinguished himself in the rodeo arena. He has carried owner* **Kristi Peterson** *to multiple World Champion Barrel Racer titles, won the NFR Barrel Racing Average three times, and been AQHA Barrel Horse of the Year three times.*

and hazed steers on him and he was a winner at all events. If all the Driftwoods were like him, I've been missing out on good horses all my life."

Goldenwood Jaye (Woody)   A capable palomino gelding that is not only a top rope horse but is bred to be be one. He is sired by Hesa Gill Ike (owned by Jim Wheatly) by White Lightning Ike by Driftwood Ike out of Driftin' Duchess by Easy Keeper and out of a Lucky Blanton daughter. His dam is by Frosty Tops who goes back to both Joe Hancock and Lucky Blanton. Frosty Tops was himself a great rodeo horse during the 1990s, ridden not only by Jim Wheatly but by many other top fifteen ropers, and made several trips to the NFR. Raised and trained by the Wheatlys, "Woody" has carried son Wade to the top fifteen standings and berths at both the 2000 and 2001 NFR, proving that "the Driftwoods do breed on."

French Flash Hawk (Bozo)   A blaze-faced bay gelding by Sun Frost by Doc's Jack Frost and out of Casey's Charm by Tiny Circus, "Bozo" traces back to Driftwood on the top side of his pedigree. Sun Frost was out of Prissy Cline, a daughter of Driftwood Ike by Driftwood. Purchased by Kristi Peterson of Elbert, Colorado for the paltry sum of $400, the gelding has carried his owner to multiple WPRA World Champion Barrel Racing titles, been the AQHA Barrel Horse of the Year three times, won the NFR average three times, and earned thousands of dollars. Among his wins Bozo can count the Cheyenne Frontier Days and Calgary, Canada—worth $50,000.

Isle Drift   A handsome black 1993 stallion owned by Gene and Angela Moench and Tyler Morris (all of Valparaiso, Indiana), by Isle Breeze by Top Breeze and out of Nikki Pine by Barry Pine/Drift's Jewel by Double Drift, his pedigree sparkles with AQHA Champion producers and champions themselves. Although unshown due to a back injury, Isle Drift was trained by Doug Williamson who said "He rode like a Cadillac." Isle Drift passes that ability on to his foals who are successfully making their ways in the show ring.

Since he has moved to Indiana, Isle Drift has introduced the midwest to the doin' ability of the Driftwoods. He is being crossed on Moench's Driftwood-bred mares to intensify the proven performance potential of the strain. His foals have traveled far from Indiana—to California, to Texas, to Colorado and to numerous states in between. Since he is a relatively young individual his offspring are

Fig. 9-24. Photo courtesy Gene Moench.

Isle Drift, *a black son of Isle Breeze/Nikki Pine by Barry Pine out of Drifts Jewel by Double Drift, is owned by Gene & Angela Moench and Tyler Morris of Valparaiso, Indiana. He is doing his share of acquainting midwest horsemen with the performance ability of the Driftwoods.*

just beginning to establish his reputation as a usin' horse sire.

While Gene Moench had been in the Quarter Horse business for some time, it has only been a few years since he learned of the Driftwoods. One look at the Jim West horses in Ione, Oregon was enough, and it wasn't long before Isle Drift took up residence at Bar None Performance Horses. Since then, he has been a strong supporter of the Driftwoods and feels that Isle Drift is an excellent example of how well they cross with other lines. "A little Driftwood does any horse a lot of good."

It is because of the performance of Driftwood's sons and daughters, as well as his grand- and great-grand-get, that his name has been kept alive. Only a truly prepotent stallion could pass on those genes through so many generations and maintain such capable performers and develop so strong a reputation. On ranches, in the show ring and the rodeo arena, areas where a contestant is only as good as his mount, the Driftwoods are still sought after and treasured. They have been there, done that, and keep coming back for more.

10

KEEPING THE FLAME ALIVE

"The presence of his name in a pedigree has become a hallmark of the qualities Quarter Horse breeders seek in establishing the highest standards of performance in their programs."
—Katy Peake

Buck Nichols
George Cline
Asbury Schell
Perry Cotton
Roy Wales
Jake Kittle
Stanley Johnson
W. C. "Tex" Oliver
Jim West
Dale Smith
Doc Whitman
Freeland Thorson
Will & Cheryl Hall
Will Gill & Sons
Mel Potter
Henry Kibler
John Balkenbush
Jim Morris
Gene Moench
Tom Eliason
Babbitt Ranches
Arnold Quarter Horses
Mike & Arlene Pixley
Perry Johns & Kenny Nichols
Amy Turner
   (Amos Turner granddaughter)
John Rudnick

Many dedicated men and women have contributed to the continuation of the Driftwood line of great working horses.

Buck Nichols—Gilbert, Arizona   One of the first horsemen to recognize the siring ability of Driftwood was this Arizona cowboy. After Buck acquired Driftwood, he trained the bay stallion as a rope horse and bred mares to him, including daughters of Clabber. After the Peakes purchased Driftwood from Asbury Schell, Nichols continued to buy colts and fillies.

The Nichols family moved to Arizona from Cheyenne, Oklahoma in 1918. Before long, they were farming and ranching at Gilbert. Since horses were an everyday part of his existence, it wasn't long before the boy was "makin' a hand"— farming, cowboying, and roping. He learned early that a mount had to have the right kind of conformation and a willing mind. He wanted a horse that could run and win, and stay sound—and then go on and make a cowhorse when the racing days were over. If they didn't fit that criteria, he didn't want them in his breeding program.

Buck Nichols was a tough hand with a rope, an early member of the Rodeo Cowboys Association (he was a PRCA Gold Card holder), and won many of the big contests in Arizona, California and New Mexico during the 1940s and '50s. He qualified for one of the early PRCA Finals in the Team Roping, but just didn't want to travel out of state to compete. Both of his sons, Don and Nick, were good ropers and horsemen as well. Grandsons Hugh, Glenn and Kenny followed the pattern, riding horses produced by their grandfather's breeding program.

While the Nichols' stallions (Driftwood, Clabber, Frosty Joe, Colonel Clyde, Wagon N, and others) were well known, Buck always felt the backbone of the program were the broodmares—and he had good ones. All were broken to saddle, raced, cowboyed on, and rodeoed. Knowing that they could perform gave him a pretty good idea of what they would produce. People saw how they worked and often requested a colt out of a particular individual after she became a broodmare. He raised good horses his entire life, receiving an award from the American Quarter Horse Association for 40 years of continous registration.

Horse operations which have been built on Buck Nichols' breeding program include Stanley Johnson, Haythorn and Babbitt Ranches, and Bob Jordon. The list goes on with the horses of numerous large and small breeders tracing back to the Nichols' foundation.

Buck realized the performance and producing potential of the Driftwood daughters early and bought or traded for many of them. Rodeo hands knew the good mares that his sons rode as well or better than the geldings. Among those individuals were Chilina, Chakaty, Lady Lux N, and Woodfern.

Buck Nichols was also an admirer of good Thoroughbreds such as Piggin' String, Ariel, Depth Charge, and Man O War. He believed that much of the good Thoroughbred blood infused in the Quarter Horse was what helped to make the breed.

The Nichols family is still raising good horses. Son Nick is at Queen Creek, Arizona; grandson Glenn at Chandler, Arizona; and grandson Kenny at Waco, Texas. And they still believe in the performance ability of the Driftwoods.

George Cline—Tonto Basin, Arizona The Clines were a family of cattlemen and ropers tracing back several generations in the rugged Tonto Basin of central Arizona. The first had settled in the country during the 1880s when the Apaches were still on the warpath. They survived raiding Indians, low cattle prices, droughts and the bloody "Pleasant Valley War" which almost depopulated the area.

Frequently, the only way to gather cattle was to rope 'em and lead them in. That took good horses and the Clines made a practice of riding the best. They would catch wild ones out in the brush and attend the scattered ropings in the area. Both George and his brother, Leck, collected more than their share of winnings at these informal affairs. When organized rodeos began they became regular competitors, although they didn't travel out

Fig. 10-1. Photo courtesy Kenny Nichols.

Buck Nichols *with one of his well-bred 2-year-olds. The filly is a full sister to grandson Kenny Nichols' Triple Drifter KN. Buck recognized the Driftwood magic early, roped on him and then bred many mares to him. The Nichols horses are a blend of Driftwood, Hancock and Clabber, along with Joe Reed II, Lucky Blanton and King.*

Fig. 10-2. Photo by George Axt; courtesy John Fincher.

Chakaty, *by Driftwood and out of Lady Lux N, was bred by Buck Nichols, she was later owned by John Fincher and Al Lauer.*

of the Sunshine State much (George won Madison Square Garden in the 1920s). They also raised horses and always had a few which could run pretty well, competing on the Arizona tracks with success.

In 1940 George noticed a bay stallion that Buck Nichols was roping calves and team tying off of and decided that he had to own him. That horse was Driftwood (Speedy). At the annual Payson rodeo one year Joe Bassett matched Tommy Rhodes on ten calves and beat him—and borrowed Speedy from Buck Nichols to do it. George Cline won $1,000 on the match and used it to buy the stallion. Both George and Leck made everyone "sit up and notice" when they ventured out from the ranch on him. They did breed the stallion to their own—and a few friends'—mares. Driftwood was already known to sire the kind that could run and look at a cow.

One of those foals was a bay stallion which was named Cowboy C (to separate him from another Driftwood son then on the short tracks called simply Cowboy and later registered as Cowboy Schell). Joe Bassett, also of Tonto Basin, bred Cowboy C but George soon purchased him. Not only did the bay stallion make a top contest mount but he sired a number of good colts for the Clines and their neighbors. During the years he was on the road, Leck Cline often heeled on him behind Asbury Schell, who was riding Cowboy Schell.

Asbury Schell, another Arizona roper and close friend, decided that he needed to cinch his saddle on Driftwood. After some haggling, George Cline sold the stallion to Schell for the same amount which he had paid for him.

The Cline family is still ranching in Arizona and raising a few good horses. And when it comes to catching a wild steer off the side of a mountain, those men can go do it.

Asbury Schell—Tempe, Arizona One of the most fiercely competitive hands ever to pick up a rope was Asbury Schell. Born in the Tonto Basin of Arizona in 1903, he got his start roping at an early age. His father, Harley Edward Schell, was a cowman of the old school and ranched in the rugged country where often the only way to catch a steer was to snare one blowing wide open off the side of some rocky mountain.

In the late 1920s, Asbury got his start rodeoing around home, riding a chunky black gelding called Midnight. It was 1936 before Schell ventured out of state, winning his first trophy buckle at a California rodeo. Because of his small size (5 foot 8 inches and 140 pounds), Asbury learned early the value of a horse which would stop and get back, giving him control over big calves. He was also a tough team roper, equally adept at either end and dallying or team tying.

Asbury Schell's ability with a rope earned him money in all three events. He was World Champion Team Roper in 1937, 1939, and 1952, although he had stopped competing full time by then. The win of which he was proudest was taking the 1939 Single Steer Roping title at Pendleton. He also won the team roping at Salinas, with John Rose as his partner, in 1954 for a whopping $3,400. During his long career, he was always exceptionally well mounted.

One of the good horses which he rode was Driftwood, purchased from his friend George Cline for $1,000. On Driftwood (whom he called Speedy), Asbury roped calves, headed, heeled, tripped steers and hazed bulldogging steers. He also bred the stallion to a limited degree, adding to the horse's reputation as a sire. In 1943, faced with gasoline rationing due to World War II and the curtailment of rodeos, he sold Driftwood to Channing and Katy Peake of Rancho Jabali, California. That part of Driftwood's history has been covered in preceding chapters.

Schell remained a Driftwood fan, purchasing and riding Cowboy for ten years and then Cherokee Jake after that. He also purchased Maestro, another Driftwood, for his son Eddie to heel on. Eddie was 1954 Team Roping Champion. Asbury pastured cattle near Coolidge and continued to rope for a number of years, always riding Driftwood horses.

Asbury Schell died at Cottonwood, Arizona in 1980.

Perry Cotton—Visalia, California A "horseman's horseman" is the term which can be applied to Perry Cotton. During his long career he has owned and bred some of the finest Quarter Horses in the nation, put together several outstanding bands of broodmares, won in the show ring and trained a number of winning Quarter Horse and Thoroughbred race horses. In his younger years, he was also a tough calf and team roper. His life has mirrored the development of the Quarter Horse industry.

Perry Cotton has owned and used such important Quarter Horse stallions as Tony, Quicksand, War Chief, Bras d'Or, Booger H., Devil Dust, and Brown Bob.

Cotton was quick to recognize King P-234 as a broodmare sire and acquired such producers as

Shu Cat, Queen Ann, Rocksea, Spiderette, High Tone, Red Jane C., and others for his and Rancho Jabali's programs. He put together what has been considered the greatest collection of Lucky Blanton daughters ever to cross with Driftwood. He also owned and stood Booger H. by King, and considered him "The best horse I ever owned."

Cotton remembered how he purchased Booger H. "I was at a sale in Texas. He came into the ring and was started at $500. Knowing that they would "nickel and dime" him until he got up to $4,000 or $5,000, I stood up and shouted $1,500. That killed 'em dead and left everyone wondering about that crazy Californian. I bought the horse."

A native of Arkansas, Cotton grew up in Oklahoma. As a child, he went to a one-room school house in northeastern Oklahoma where he was two grades above Ben Johnson, the movie actor and champion roper. In 1933 he sold a 3-year-old Quarter Horse filly (called Steel Dust) for $35. It took, in his 1962 words, "A pretty fair filly to bring that—the same as $2,000 today." With that money he made his way to California and eventually into cotton farming in the San Joaquin Valley. He raised cattle and wheat on land near Corcoran. And he continued in the horse business, developing a successful breeding operation at Visalia.

Realizing that the Quarter Horse was going to be "big business," Cotton began to develop a breeding program even before the AQHA was organized. One of his early stallions was Brown Bob, who he roped on and bred before selling him to a breeder in the Los Angeles area.

At the time he decided to enter the Quarter Horse business, Texas was the source and Cotton studied the various bloodlines available there. If he was going to raise horses, he wanted the best. In his opinion, Jess Hankins, the Wardlaws, and Duaine Hughes had the best horses available. He attended the 1941 Duaine Hughes sale at Big Lake, Texas. He purchased Devil Dust 1088, a young stallion who he developed into a solid calf roping mount. Devil Dust was later sold to Gordon Davis who campaigned him on the professional circuit. His next owner was Chuck Sheppard, one of the better ropers and horsemen to ever nod for a calf.

In 1947 Cotton and the Peakes formed a partnership to raise Quarter Horses, utilizing both Driftwood

Fig. 10-3. Photo courtesy Peake files.

Perry Cotton *of Visalia, California accepting a ribbon for* **Quicksand**.

(owned by the Peakes) and the Red Dog and King stallions Cotton had. The results from that plan produced good using prospects and potential broodmares that found ready buyers. That partnership lasted until 1962 when the horses were dispersed.

Cotton was not only a supporter of Driftwood, regretting that he had never tried to purchase him, but an admirer of Katy Peake. "She was a wonderful person, a born horsewoman with an inborn love of them, and I should have listened to her opinions more," he remembered.

As a roper and usin' horse man, Perry Cotton was well aware of the performance ability of the Lucky Blantons. He began to collect as many top daughters of the stallion as possible. Katy Peake started the program by giving him Dusky Ruth and Baja Fiji and then sold him several more. She also offered him the use of Driftwood—an offer he didn't take and admits was a mistake. When he realized what producers those Lucky Blanton mares were, Perry "wore out a Cadillac putting more of them together. With Driftwood and those Lucky Blanton daughters, I would have produced the greatest set of working horses in the world and been set for life," recollects Cotton. Those mares were sold at his dispersal sale in 1962.

After the closing out of the Peake/Cotton partnership, Perry continued in the horse business. Always a devotee of speed, he tailored his operation towards producing runners. One of the stallions he purchased was Lucky Bar, a double-bred son of Three Bars TB. One of the foals which he bred during this period was Impressive, a great halter horse who has had a terrific impact upon the halter segment of the Quarter Horse world. (He sold the mare before the sorrel colt was foaled.)

Fig. 10-4. Photo courtesy Peake files.

Roy Wales *of Queen Creek, Arizona. He became acquainted with Driftwood shortly after Buck Nichols purchased the stallion. Over the years he bred mares to him, raised good horses and stood Driftwood Ike.*

Fig. 10-5. Photo by Livingston.

Ozzie & Judy Gillium, *long-time Driftwood breeders. Judy is the daughter of Roy Wales.*

The relationship between Perry Cotton and Katy Peake continued even after the partnership was dissolved. When Rancho Jabali dispersed in 1966, Perry was asked to inspect all the horses before the sale by Max Schott. Max knew of no one from whom he would get such an honest evaluation as Perry Cotton or who had been so closely associated with the program and the goals which had been set over twenty five years before.

Cotton eventually phased out all of his Quarter Horses and concentrated upon Thoroughbreds, raising foals and training horses at the track. As always, his efforts in any endeavor with horses has been successful and he has produced many winners, foals with high potential, and quality breeding stock. Among the stakes winners he has bred and raced are Liz Tasto, winner of $110,425; Ali Kato, winner of $182,698; Cotton Bloomers, winner of $292,060; Dooley; and scores of California-breds.

At 90-plus years, Perry Cotton is still in the horse business. He does admit, though, "One thing I've learned about the horse business is that most successes come from sheer, blind luck." In his case that "luck" has to be combined with an astute eye for a horse and a knowledge of what a combination of bloodlines will produce.

Roy Wales—Queen Creek, Arizona This cotton farmer, horseman, and roper was introduced to Driftwood early in the story. He first saw the stallion when he was owned by the Nichols family. The same year (about 1939) he roped on him at the Prescott rodeo. From then on, he was a Driftwood fan and bred many mares to him, both in Arizona and after the stallion went to California.

After the Peakes purchased Driftwood, Roy Wales left three mares at Rancho Jabali permanently. He paid $365 annually for both care and breeding. "That's what Katy wanted and that is what I paid," remembered the horseman. When the foals were weaned, he would take them to Arizona, halter break them, and then turn them out on a half-section of desert pasture. The following year, the colts came in for a refresher course and then went back to pasture. At two, they came in again and were started under saddle.

Roy Patton did most of the early work on those colts, assisted by Roy's oldest daughter, Jerry.

When the colts were ready, they went to the roping arena where Roy Wales, Roy and Jess Patton, Walt and Don Nichols, John Rogers, Abe Logue, and B. J. Pierce all played a big part in their education. During the 1950s and '60s, Wales kept 100 calves, 50 roping steers and five to ten horses in training all the time.

Pierce remembered Roy and Tom Finley sitting, whittling, and watching him rope on the colts. When he was done for the day, both men would get up and leave the arena without saying a word. B. J. commented that Roy Wales was gifted at getting a horse ready for a show or the rodeo arena. "If Roy told you to go rope a few on this horse, the horse would feel like he was just made for you." He would not sell a horse if he didn't feel that it was right for the person who wanted it.

Roy Wales was quite a roper himself, both on calves and steers. His biggest roping thrill, however, was in the arena at home with his children, matching for red soda pops.

Among the good Driftwoods that Wales owned were Driftwood Ike, Firewood, Hallie Wood, and Wooden Nugget. Those horses were not only winners in the show ring but the kind that a contest roper could make money on. B. J. Pierce hauled Wooden Nugget one summer and fall, making the big rodeos and matched ropings on him. Woody was not only a good rope horse but also carried Walt Mason to numerous wins as a stock horse. He was later crippled and retired from the arena.

Of all the Driftwoods which he owned, Roy Wales is best remembered for Driftwood Ike. He purchased the dun stallion as a 2-year-old in 1956. Driftwood Ike was not only a top contest mount, with many of the tough ropers of the period riding him at Arizona rodeos, but a prolific sire. To many individuals, he was considered to be Driftwood's best siring son. His foals were a rodeo cowboy's dream and many of the top hands rode them. Often, the National Finals contest was a showcase of "Driftwood Ikes." One year there were 18 of them there—the next year 21.

In the 1990s many of the best ropers in the country were still riding Driftwoods. Spui Drifter with Clay O'Brien Cooper in the saddle, Cisco Ike with Brian Anderson, and Luca Lulu with Rube Woolsey—all by Mr. Bar Truckle and out of Driftwood Ike mares—made their appearances at the NFR.

Roy Wales died of cancer in the 1990s, but his children are still carrying on the love of the Driftwood legacy. Jerry is ranching and raising horses, Jimbo is farming as well as raising and training Driftwoods, and Judy is raising horses by Mr. Bar Truckle and Driftwood mares. And like their father, they believe that "if a horse has two drops of Driftwood blood, he's got to be good."

Jake Kittle—Patagonia, Arizona One of the early horsemen to recognize the value of Driftwood daughters was Jake Kittle. In 1952, he, the Espils, and Bob Lockett leased Red Man (by Joe Hancock) from Kenneth Gunter of Cochise, Arizona. The roan stallion had been a successful race horse, a top rodeo mount and a very capable sire of performance foals from often average mares. Kittle's first mare purchase was O-See-O, by Driftwood out of Little Moore TB and in foal to Booger H., from the western artist, Joe de Yong. O-See-O was twice a Grand Champion Reining mare in California. Kittle commented about O-See-O, "She had the best head I ever saw on a mare." He also acquired four two-year-old Driftwood daughters—Driftwood Katy, Miss Linwood, Swiftwood, and Brown Beulah from Rancho Jabali.

Breeding Miss Linwood to Red Man produced Cibecue Roan. Under the handling of Grady Stewart, the versatile roan stallion became an AQHA Champion Working Cow Horse and World Champion Team Roping Horse. Swiftwood, bred to Red Man, foaled Redwood Man, another show winner in Working Cow Horse and Roping with John Hoyt in the saddle.

Not wanting to breed Cibecue Roan to his own daughters, Kittle purchased a group of mares from Mavis Peavy of Colorado. These mares were strictly old-line foundation bred, going back to Red Dog, Ding Bob, and Saladin for the most part. The results were good. One foal became the AQHA Champion El Campanero and another was Thrifty Miss. Others included Peavy's Breeze, Nikki Tea, Peavy's Puddin, Bella's Queen, Jillaro, and Margie Ding.

Jake Kittle carefully planned his breeding years in advance and it paid off with mares which were used in programs throughout the West and usin' geldings on many ranches. The Driftwood

Fig. 10-6. Photo by Willard H. Porter.

Jake Kittle and Red Man. *Kittle utilized the Driftwood/Hancock cross very successfully. Among the Driftwood daughters which he owned were O-See-O, Driftwood Katy, Miss Linwood, Swiftwood and Brown Beulah.*

daughters produced numerous good sons and daughters. The Peavy mares, bred to Cibecue Roan and other stallions, were also good mothers. Fred and Clara Wilson of Newcastle, Wyoming used the offspring of Kittle's Driftwood daughters in their mare line, adding Hancock, Dash for Cash, and Doc Bar breeding on the top of the pedigree. Arizona breeders utilized Redwood Jake, Cibecue Roan and like-bred stallions to carry on the Driftwood/Joe Hancock combination that has so appealed to horsemen that value a horse for what he can do.

Jake Kittle is now retired to a small place in Patagonia, Arizona.

Stanley Johnson—Miller, South Dakota A rancher from South Dakota, Stanley Johnson learned the value of a good horse at an early age. Not content with "just horses," his first venture into the registered business was the purchase of Poco Speedy, a son of Poco Bueno, who he showed to an AQHA Championship.

In the late 1950s, Johnson and his family began to visit Arizona during the winters. There, he was introduced to the Driftwoods. He purchased daughters of Driftwood and Driftwood Ike, crossing them with Poco Speedy. The daughters of Poco Speedy and Driftwood-bred mares were the foundation of the Johnson breeding program.

Then he purchased three sons of Driftwood Ike from Roy Wales. Of the three, Stanley and his son Randy kept Orphan Drift. He was crossed on the Poco Speedy and Driftwood mares. There was soon a great demand for those foals.

Johnson's next step was to acquire Doc's Jack Frost, by Doc Bar and out of the race mare Chantela. Bred to Orphan Drift daughters, this stallion sired offspring which became world champion qualifiers and world champions.

Stanley's last stallion was Sak 'em San, a son of Peppy San. He was able to see only two of Sak 'em's foal crops before passing away in 1982.

Stanley Johnson was one of the earlier breeders to recognize the quality of the Driftwood influence, and worked hard to preserve it. He used Driftwood-bred mares as the base, bringing in outside blood as he felt it was needed. His influence is still felt today with numerous breeders basing their programs on what he began.

People who were fortunate enough to have seen Stanley's outstanding band of mares still talk about the consistency and quality. There were very few herds in the country which could compare.

Randy, Stanley Johnson's son, is in the Quarter Horse business and still raises Driftwoods at Cave Creek, Arizona. He is standing Driftwood's Jo Jo, a son of Orphan Drift out of Jo Jo Wood by Poco Speedy, carrying on the proven combination of bloodlines which his father established.

W. C. "Tex" Oliver Tex Oliver was a rancher, a horseman and a breeder of good horses. He was quietly confident, knowledgable, skillful with horses and cattle and well thought of. He knew what he liked in a performance horse and his program never varied much. Early in his career he discovered the Driftwood daughters and used them extensively. As a ranch manager he moved around, from Santa Barbara County, California to Nevada to Los Banos and then back to Nevada. In each move, he took his horses with him.

He recognized that the Driftwood/Hancock cross produced a superior performance horse and seldom deviated from that. When Katy Peake let him use War Chief (considered the fastest son of Joe Hancock), he continued to follow that line.

Crossing War Chief with Brown Beulah by Driftwood out of Queen Ann (by King), was a master stroke—producing both War Drift and War Concho. He also bred Brown Beulah to Red Man, producing Redwood Jake and May Belline who, when crossed with Speedywood, foaled Speedy Man.

According to those who knew him, Tex Oliver planned several generations ahead and knew what he wanted to raise. His ability to combine bloodlines and to assess which individuals should be crossed resulted in horses that were popular with cowboys, ranchers and rodeo hands. The results of his breeding program has influenced numerous breeders, including the Haythorn Ranch, Charlie Judd, Henry Kibler, and John Balkenbush, to name a few.

Dale Smith—Chandler, Arizona The scion of a pioneer Arizona ranching family, Dale Smith has been a rancher, roper and horseman all of his life and been a winner in every endeavor. Considerred by many to be one of the most complete ropers (he could win roping calves, heading, heeling or tripping steers) of all time, Smith earned a total of 20 NFR back numbers, including being the first man to qualify in all three roping events in one year, and won most of the major rodeos in the nation during his long career.

Dale Smith realized early that to win he had to be mounted and always rode some of the best horses in the business. As an Arizonian it was natural that he turn to the Driftwoods. Poker Chip Peake, a son of Driftwood and considered the greatest calf horse to ever step into the arena, was one of Dale's mounts. Another was Shady, a black Driftwood grandson who could catch calves, head, heel or trip steers. Speedywood, by Driftwood, could get the job done in the arena and was also considered an outstanding sire of rodeo horses.

As a successful cattle rancher, alfalfa and cotton farmer, Smith recognized the need for good horseflesh and bred Speedywood to his select band of mares. That move not only gave him top ranch horses but put a number of tough ropers a-horseback. A number of the Speedywood sons, and the men who rode them, are mentioned in this book.

Even after Dale Smith curtailed his rodeo activities he continued to rope. For many years he competed

Fig. 10-7. Photo courtesy Peake files.

Tex & Inez Oliver with **Woodtick**. *Tex Oliver bred many good Driftwoods and War Chiefs.*

Fig. 10-8. Photo courtesy Hoofs and Horns magazine.

Dale Smith of Chandler, Arizona. World Champion Roper, rancher and horseman. He owned and roped on Poker Chip Peake, Shady, Speedywood and a number of other Driftwoods. As a breeder he stood Speedywood for several years.

at the big team roping jackpots and always won his share riding a Driftwood-bred horse.

He was not only a winner in the rodeo arena, but worked hard for the betterment of the sport and contestants. He served fourteen years on the PRCA Board of Directors, including twelve terms as President. With his level-headed business approach he guided the organization through some rough times.

While he continued to ranch, Smith later directed his horse breeding activities to the Thoroughbred racing track. As might be expected, he was also successful in that venture.

Doc Whitman/Freeland Thorson/Will & Cheryl Hall—Paradise Valley, Nevada The story of three generations of Driftwood breeders began when Freeland Thorson transferred to Cal Poly from the University of Idaho. As a roper he was immediately impressed with the ability of the Driftwood horses he saw West Coast ropers riding. Thorson and Bill Gibford, head of the college horse department and a great supporter of the Driftwood horses, had taken a college-owned mare to Candy Spots, a Thoroughbred stallion owned by Rex Ellsworth. On the return trip, they stopped at Rancho Jabali. Katy Peake asked Gibford to train a two-year-old for her. The horse selected was Double Drift.

A year or so later, when Catherine Peake was leaving the horse business, Thorson's father-in-law, Dr. J. R. Whitman, purchased the stallion. Whitman had Double Drift shown in both halter and performance. A versatile individual, the black stallion won points in Heading, Heeling, Barrel Racing, Cutting and Western Pleasure. A pro-caliber rope horse, Double Drift was also competed on at many of the big team ropings on the Pacific Coast.

In 1967, having sold his ranch at Lodi, California, Doc Whitman gave the horse to his daughter, Diana. The Thorsons took Double Drift to the Idaho ranch. Thorson daughter Cheryl trained and showed numerous champions throughout the inter-mountain area, as well as successfully competing in open and intercollegiate rodeos, all with Double Drift foals.

The Driftwood program continues today with Cheryl Hall and her husband Will, raising horses at their Bar 7X Ranch at Paradise Valley, Nevada. They are standing Dynaflow Drift, a son of Double Drift out of a Country Boy Lauro mare, and Doc Whitman, by Mr. San Peppy and out of a Doc Bar daughter.

Fig. 10-9. Photo courtesy Foundaton Quarter Horse Magazine.

Doc Whitman and Double Drift. *Whitman purchased the stallion from Katy Peake, showed and roped on him, and then gave him to his daughter, Diana Thorson.*

Fig. 10-10. Photo courtesy Cheryl Hall.

Cheryl Hall, *Bar 7X Ranch, Paradise Valley, Nevada with* **Dynaflow Drift,** *a son of Double Drift out of a Country Boy Lauro mare. Mrs. Hall, daughter of Freeland and Diane Thorson, is carrying on the Double Drift line of horses started by her grandfather Doc Whitman.*

Will Gill & Sons—Madera, California The Gill family has ranched in California since 1873, when Levi Gill left Iowa and brought his family to Poterville. Among his ten children were the twin sons, Fred and Will. In 1902 they formed a ranching partnership, and by 1921 had accumulated thousands of acres of grazing land.

Fred formed the Gill Cattle Company and eventually operated in Arizona, Oregon, Wyoming, and California. His sons, Adolph and Emmett, really took an interest in the quality of the ranch horses and used both Mark and Bear Hug as sires. During the 1960s a son of Three Bars, Bar Tonto, was also in use.

Will eventually developed the partnership of Will Gill & Sons, ranching in Madera, Merced, and Santa Clara Counties in addition to retaining the home ranch in Tulare County. The operation handled some 7,000 stocker cattle annually. His son, Ralph, stayed on the home ranch while Ernest managed ranches at Madera and Gustine. Will's son-in-law, Clay Thompson, ran the feedlot in addition to establishing Thompson and Gill Manufacturing, which fabricated cattle handling equipment.

Although Will Gill & Sons had been raising ranch horses for years, the registered program didn't begin until 1947. Will Gill, Jr., recently out of the Army, acquired the 2-year-old stallion Easy Keeper P-12044 by Driftwood and out of Smoky McCue. An injury at an early age curtailed his performance career, but he was a successful sire. To complement Easy Keeper, a truck load of registered mares carrying the blood of Peter McCue, Joe Hancock, Chief, Texas B., Tommy Clegg, Snicker, and Joe Tom was purchased from W. M. Howard.

As proof of the popularity of Easy Keeper foals among the ropers, the stallion was honored as the leading sire of rope horses at the 1958 California Championship Team Roping at Oakdale. His colts were easily recognizable by their conformation and most carried his bay color.

The next step was to purchase Pelican P-55093, by Joe Hancock Jr. Pelican was a former World Champion Quarter Running Stallion. A Pelican daughter, Aliso Gill 3, produced four AQHA Champions and is the granddam of two-time Super Horse Rugged Lark. Crossing Pelican on Easy Keeper mares produced many ranch and arena horses.

To intensify the Driftwood influence, White Lightning Ike by Driftwood Ike and out of Katy Was A Lady by K4 Hickory Skip was purchased. The buckskin stallion has the traditional good disposition, cow sense, athletic ability, and natural stop of the Driftwoods, and passes those traits on to his foals.

Ernest Gill was World Champion Team Roper in 1945 and Will, Jr. was another tough roper. The two men have won many big ropings, including the Chowchilla Stampede and the Oakdale 10-steer. Will Jr.'s son, David, has been an NFR Team Roping qualifier. The Gills knew the value of good horses for both work and play, and have always kept themselves and their cowboys well mounted. In fact, keeping good horses under their cowboys has been a major factor in maintaining a stable

Fig. 10-11. Photo by Livingston.

Will Gill, Jr. *and son* **David** *of Madera, California. The Gill family has bred many outstanding Driftwood horses over the years, using Easy Keeper and White Lightning Ike.*

Fig. 10-12. Photo by Livingston.

Bennie Norman *of Madera, California. Norman has been long-time manager of the Will Gill & Sons horse operation in addition to being involved in Norman-Pentorali Quarter Horses.*

workforce. The horse program has been a continuing source of revenue for the Gills as ranchers and cowboys are always ready to purchase a good prospect.

The Gill family is quick to give credit to Bennie Norman, manager of the horse operation. "Bennie has been with us for many years. He virtually lives with the horses and is responsible for much of the success of the operation," says Will Gill, Jr.

It has been said that more World Class team ropers have competed at the National Finals Rodeo on Gill-bred horses than those from any other ranch. Some of the horses that have carried contestants are Pellet, Gill, Snickerton, Easy Gran (Cadillac), Easy Doc Sox (Buddy), Easy Chaw (Sherman), Gold Cloud Miss, Buckshot Ike, Off Limits Ike, Madera Gill Ike, Booger Ike, Lightning Storm Ike, Blue Light Ike, and Frostys Tops.

Among the team ropers who have ridden Gill-bred horses are 4-time World Champion Jim Rodriquez, Jr.; 2-time World Champion John Miller; multi-time Finalist Doyle Gellerman; 7-time World Champion Jake Barnes; World Champion David Motes; Gary Walker; Ron Goodrich; Jim Peterson; David Gill; and Liddon and Cody Cowden.

The 2001 Will Gill and Sons Sale, the first in many years, saw 60 head of White Lightning Ike and other Driftwood-influenced foals and older horses go through the ring at an impressive $3,690 average to an audience of enthusiastic buyers. That proved that a 50-year breeding program geared to produce outstanding performance horses was on the right track.

While Will Gill & Sons set out to breed ranch horses, the operation has put scores of top rodeo cowboys a-horseback for several generations.

Mel Potter—Marana, Arizona Like a number of Driftwood enthusiasts, Mel Potter was introduced to the Driftwoods in the rodeo arena. His family moved to Arizona when he was 9 and he's had a rope in his hand ever since. He attended the University of Arizona, rodeoed for them and went on to compete at the pro level. He qualified in the

Fig. 10-13. Photo by Jim Morris.

Mel Potter *of Marana, Arizona, and* **Frenchman's Hayday.** *Potter is a long-time roper, Driftwood fan and breeder who rode many outstanding Driftwood sons and grandsons during his career in the arena.*

top 15 in the calf roping at the first National Finals Rodeo in 1959. During those years he discovered that 40 to 50 percent of the rope horses in the Southwest were Driftwoods.

While in college he became friends with John Kieckhefer, who owned and hauled Driftwood Ike at the time. A few years later, Mel, John, and Chuck Sheppard (John's father-in-law) decided to go into the breeding business with Driftwood Ike daughters as the base of the operation. Over the years, Potter has expanded the operation, but the majority of his mares are still Driftwood Ike-bred.

Potter currently keeps around 20 mares and breeds the majority of them to Lone Drifter, a son of Driftwood Ike whom he purchased from

Fig. 10-14. Photo by Jim Morris.

Broodmares and foals at the Potter Ranch *near Marana, Arizona. The breeding program preserves and promotes the Driftwood bloodline.*

Kieckhefer and Sheppard as a three-year-old. He has also loaned the dun stallion to friends, including the Haythorn Ranch at Arthur, Nebraska.

The foals from this operation sell rapidly to ropers and usin' horse folks who want to be well-mounted in the years to come. Of the Lone Drifters which he's raised, he says, "They have excellent minds, they can run, have lots of cow, are natural stoppers and they 'make' so easy."

Henry Kibler—Chandler, Arizona Another Arizonian who got the Driftwood bug from the roping pen is Henry Kibler. His grandfather settled in the Salt River Valley near Phoenix in 1906, established a dairy, and the family has been there ever since. His sons mark the fourth generation in the business. Like many of the farmers and ranchers in the area, Henry's father roped and it was natural that the son did as well. He remembers roping at the nearby Glenn Motts arena when he was just seven years old. (Motts qualified for at least one NFR in the team roping, his sons roped at the pro level, and son David was a World Champion.) Many of the men Henry and his father competed against were mounted on sons and grandsons of Driftwood and he was impressed by how well the horses took to the arena.

Kibler kept a few mares, breeding them to Driftwood Ike, Speedy II, Chipperwood and Pattonwood (a son of Chipperwood which he leased for several years). About 25 years ago he purchased Speedyman from Harper McFarland (see Speedyman bio) and really got serious about breeding Driftwoods. He and his wife, Nancy, realized that the Driftwood blood was slipping away and decided to concentrate on developing a high percentage breeding program before all the Driftwood grandaughters were gone. Kibler's next step was to visit Tex Oliver. There, in partnership with Randy Johnson, he acquired Mahala Hancock as well as Beulah's Annie for himself. Both were daughters of Brown Beulah by Driftwood and by War Chief by Joe Hancock.

Since 1980 the focus of the Kiblers has been the Driftwoods. He currently keeps 15 to 20 high-percentage mares which carry the blood of Speedyman, Pattonwood, Senor Driftwood, War Chief, Lone Drifter, Driftwood Ike, Speedy II, Easy Keeper and War Concho. They are good, solid, typy mares with the conformation and strength of

Fig. 10-15. Photo by Frank Milne; courtesy Mel Potter.

Mel Potter and Poker Chip Peake at Red Bluff, California in 1961. Dale Smith, who owned the horse, won 1st in both go rounds and Potter won second in both go rounds and the average, making it a profitable weekend for the gray gelding.

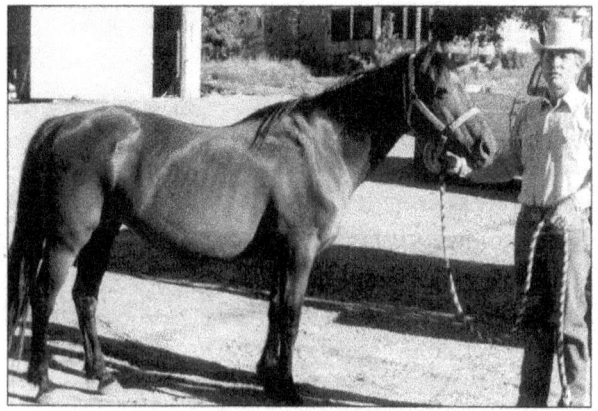

Fig. 10-16. Photo courtesy Henry Kibler.

Henry Kibler, Chandler, Arizona, pictured in 1985 at the Fuchs Ranch in Santa Rosa, New Mexico with **Polly Peake** by Driftwood. Henry and his wife, Nancy, were some of the first breeders to realize that the Driftwood line was slipping away and concentrated on maintaining it.

genetics that produces arena and ranch horses a person is proud to ride.

The four stallions on the Kibler ranch, all strong in Driftwood genetics, are Speedy Roan Man by Speedyman out of a Quarter Hancock daughter;

Fig. 10-17. Photo courtesy Henry Kibler.

Speedy Roan Man, *a son of Speedyman out of Quarter Sissy by Quarter Hancock. He is one of the stallions which Henry Kibler is currently using on his Driftwood-bred band of broodmares.*

Fig. 10-18. Photo courtesy Henry Kibler.

Two Driftwoods doing what they do best at Henry Kibler's Chandler, Arizona arena. **Sheldon Hall** *heading on* **Cortano** *(by Speedyman out of a Speedy II daughter) with* **Sterling Udall** *heeling on* **Easy Man** *(by Speedyman out of an Easy Keeper daughter).*

War Train by War Concho out of a Wayward Ike daughter; Santa Cruz Drifter by Dynaflow Drift out of a Speedyman daughter; and Driftup Speedy by Dynaflow Drift out of a Pattonwood mare. He is also adding a touch of Lucky Blanton as an outcross, blending the bloodlines of another top performance producer with a high percentage of Driftwood.

Henry Kibler is located in the "heart of Driftwood country" and has more than done his part to preserve the legacy of good horses that can do something in the arena.

John Balkenbush—Conrad, Montana "I picked up the horse bug as a kid and never got over it," is the way John Balkenbush describes himself. He grew up in Wyoming, wrangling dudes for his grandparents. Most of the really top rodeo or ranch horses he saw were Hancocks—the result of King Merritt's breeding program at Federal. He decided that when he was able to raise horses they would be that type and quality. Eventually, Balkenbush was able to put together a band of Hancock-bred mares.

A decade later he felt that he needed to refine his horses, "develop more athletic ability, make them smoother to ride and give them a smoother overall look." He chose the Driftwoods. After visiting Charlie Judd in New Mexico he purchased Guatemala Canal and several of his daughters. Guatemala Canal was by War Concho (War Chief/Brown Beulah) out of Miss Canal (Frosty Joe/Chiquita Bonnet by Cherokee Jake), combining both Driftwood and Joe Hancock bloodlines. He used him for five years with a great deal of success, producing "the kind of horse I like."

"I've been blessed several times in my life and Guatemala Canal was sure one of them," commented Balkenbush.

The Sunshine Ranch is dedicated to breeding "The Best Ranch Horses in Montana," the kind that a person can cowboy on all day long. "We get our horses behind a cow at every opportunity, go to brandings and really cowboy on 'em," said Balkenbush. "The geldings go to cowboys and ropers and most of the fillies are out on ranches— once they get there, they stay."

Currently the Sunshine Ranch is standing four stallions, all strong in the Driftwood-Joe Hancock cross. The senior stallion is Quatro Drift by Drifts Chip (Double Drift/Diamond Isle by Diamond Chip) out of Conchos Lady 010 (War Concho/Miss Canal). There is also Ikes Bar Drifter by Dynaflow Drift (Double Drift) out of Ikes Bar Girl (White Lightning Ike/Sandys Bar Girl by Son of Bar Girl); Rapid Drift by AP Frosty Knight (Sun Frost/AP Knight) out of Conchos Lady 004 (War Concho/Speedy Beth by Driftwood Ike); and Coconino Cowboy by Cowboy Drift (Orphan Drift/Poco Judy Sue by Poco Speedy) out of CO Bar Six (Deck of Wood by Speedy II/CO Bar Three). Balkenbush also used Lindsay Peake (Son of Bar Girl/Sandy Peake by Speedy Peake) for one season.

John Balkenbush has made good his vow to raise good horses and in the process put a lot of Montana cowboys a-horseback.

Jim Morris—Exeter, California A native Californian who grew up around horses, both draft and saddle stock, Morris appreciates what it takes in both blood lines and conformation to produce a good one. Morris bred his first mare to a neighboring Remount Service stallion when he was just 13 years old, paying the stud fee out of his hard-earned farm wages. The fact that he lost the colt a year later didn't deter him from making a life-long career in the horse business.

As a horseman (except for a 3 1/2 year stint during World War II when he flew 52 bomber missions) Morris cowboyed, rodeoed, ran steers and owned a saddle shop in neighboring Visalia. During the late 1940s and '50s he was aware of the impact that Driftwood was making on the using horses of the Pacific Coast and visited Rancho Jabali several times to see both Driftwood and Speedy Peake. He decided that it was time to get back into the breeding business. Impressed by the King-bred mares that the Peakes and Perry Cotton were using, the first stallion Morris purchased was Old Granddad, a full brother to Poco Bueno. In the early 1960s he sold out his breeding operation because of the press of business.

In 1987, after retirement from business and hauling his daughter, Michele, to youth shows, he saw Lindsay Peake at neighbor Russell Keeley's place and "had to have him." The good looking brown stud colt combined the modern bloodlines of Son of Bar Girl (Son O Sugar/Doc's Bar Girl) with the traditional working ability of Driftwood, King, Blue Rock and Pretty Buck. He then got busy and hunted down five granddaughters of Old Granddad, knowing that the King blood would add a solid base to his program. He also acquired two intensely-bred Driftwood fillies, Lotta Driftwood and Speedy Lil, from Henry Kibler of Chandler, Arizona. Morris was back in the breeding business. Lindsay Peake, crossed on these mares, has produced the kind of horses that a person is proud to cinch a saddle on. The brown stallion has also stood in Indiana, at Gene Moench's ranch, in Montana with John Balkenbush's War Concho mares, and in Arizona at Babbitt Ranches.

In carrying on the Driftwood tradition, Morris has been instrumental in the founding of the informal Driftwood Breeders Association and produced the Driftwood Breeders Newsletter. And without his

Fig. 10-19. Photo by Livingston.

Jim Morris *of Exeter, California, and* ***Lindsay Peake****. Morris is a long-time Driftwood breeder and one of the founders of the Driftwood Breeders Association.*

invaluable assistance, cooperation and enthusiasm, this book would not have been produced.

Gene Moench—Valparaiso, Indiana This long-time midwest Quarter Horse breeder stumbled on the Driftwoods in a unique way. During the winter of 1992-93 he saw a television program on Duff Severe, an Oregon rawhide braider, saddle maker and horseman. It featured a herd of good looking horses owned by Jim West of Ione, Oregon, crossing a creek. Those horses really impressed Moench. After several phone calls tracing down where they were, Gene found himself driving to Oregon to visit West and look at his horses.

The Moenchs were awestruck by the coal black stallion Drifts Chip, the brood mares and ranch geldings. Before leaving Gene purchased several of the stallion's daughters. A year later he returned, with Brian Wright of Chicago, and acquired another Drifts Chip daughter. He sent her to Will Gill & Sons of Madera, California, where she was bred to White Lightning Ike. The resulting foal was Ike's Double Drift. He also made arrangements to ship several of his own mares to West to be bred to Drifts Chip.

After seeing the West horses, admiring their functional conformation, athletic ability and willing minds, Moench was hooked on the Driftwoods and decided to infuse that line into his breeding program. He also began to research the history behind the strain of horses, and discovered that it had been one of the best kept secrets in the horse industry. Only ranchers and rodeo hands knew much about the Driftwoods and they were more interested in using than showing them.

Moench began a search for more Driftwood blood. With Tyler Morris, he purchased Isle Drift, by Isle Breeze and out of a Poco Pine/Double Drift mare. In 1994 the search led to Diane Kem of Deer Island Quarter Horses, breeder of the dams of both Isle Breeze and Drifts Chip. In the initial telephone conversation Ms. Kem told Gene that there was something about the Driftwoods that he needed to know but that she couldn't remember it. Nearly 30 minutes into the discussion she suddenly interrupted with "I know what it was! There never were enough of them!" She also told Gene to call Perry Cotton, long-time Driftwood fan, who referred him to Will Gill, Jr., who owned White Lightning Ike, a son of Driftwood Ike. A trip to California was planned as a result.

The Indiana horseman also responded to Henry Kibler's *Quarter Horse Journal* ad, asked about the Driftwoods and mentioned his upcoming trip to the Pacific Coast. Kibler suggested that Moench make an effort to meet Jim Morris while he was in the country, which he did. From Exeter, he went to the Gill's Madera ranch. After visiting Will Gill, Jr., Moench added a trailer load of White Lightning Ikes to his herd. One mare was sent to Jim West to be bred to Drifts Chip. That move produced Driftwood N Drift, now owned by Tracie Langley.

The California trip resulted in a close friendship with Morris, a "fountain of information" on the old bloodlines and the breeder who owned Lindsay Peake, a grandson of Speedy Peake. Morris and Moench were to partner on a number of horses.

After sending a White Lightning Ike mare to be bred to Drifts Chip, Moench responded to Jim West's suggestion to develop a network of Driftwood enthusiasts. He took the first steps to develop the Driftwood Breeders Association, an informal group of breeders. He enlisted Brian Wright and Jim Morris to help with the endeavor. The first step was an ad in the *Quarter Horse Journal* asking for information on the Driftwoods. In answer came numerous letters and phone calls from horsemen who had owned and admired the horses for years. Working with Wright and Morris, Moench developed a questionnaire which was sent out to breeders asking for more information and an indication of interest in forming a breeders' group. A mailing list of over 100 names was soon compiled. Those efforts resulted in the only "official" meeting of the Driftwood Breeders Group, held in Fresno, California in 1995, and the program was on its way. Jim Morris took responsibility for a quarterly newsletter which was mailed to the members. This newsletter compiled information regarding the great Driftwoods from the past, photographs, early breeders, and gave current Driftwood fans a rallying point. The resulting interest in what was once a "forgotten" family within the Quarter Horse breed began to boom in popularity as more and more people found out about them.

Moench's own breeding program has continued to grow and prosper. Currently he is breeding mares to Ike's Double Drift, standing at Turner's Driftwood Ranch in Denton, Texas; his Driftwood mares to Isle Drift and Smoke Shak and to his young outcross stallions by Doc's Hickory and Doc's Stylish Oak.

NOTE: While the Driftwoods have been known and kept alive by ranchers, ropers and usin' horse folks, the current popularity can be traced to the initial efforts of Gene Moench. After much searching, he added the qualities of solid conformation, athletic ability, and proven performance bloodlines to his own program. He realized that the scattered Driftwood breeders around America needed to be unified and the story told to the rank and file of horsemen.

Tom Eliason—Gregory, South Dakota Another breeder who got the "Driftwood bug" from the roping arena is Tom Eliason. While at college, he roomed with John Fincher, 1964 Intercollegiate Calf Roping Champion. Fincher owned and roped off of a son of Driftwood. Later, he owned Chakaty and Chilina, full sister to Driftwood Ike, in partnership with Donnie Nichols.

"John was high on the Driftwood horses and I was really impressed with the ones he rode," remembers Eliason. "He was the one who told me that the Driftwoods were so easy to train, had a natural stop and cow sense. Over the years, I've found these Driftwood traits seem to run truer than other bloodlines."

During the 1960s and '70s, Eliason trained horses professionally for Quarter Horse breeders, including Howard Pitzer. Tom showed the well-known Two Eyed Jack to his first reining win.

In 1967 a friend, Herb Ulmer, offered to loan him a Stanley Johnson stallion for a rope horse. Tom was willing to try the horse, a son of Poco Speedy out of Judy Sue by Driftwood. Randy Johnson had roped a few calves on the horse but he was a long way from being a finished rodeo mount.

Eliason had only ridden the brown stallion twice when he took him to an AQHA show at St. Paul, Minnesota. Tom and his Driftwood-bred mount cleaned up, winning third in Calf Roping, fourth in Working Cow Horse and tying for first with the horse that was leading the National Reining standings at the time. "Ol' Brown could really gather and stop on that bad ground," remembered Eliason with a laugh.

In 1977 Eliason and his wife JoAnn settled in Gregory, South Dakota. He entered the insurance business but didn't hang up his ropes, competing at nearby rodeos and jackpots.

In 1983 Tom located a double-bred Driftwood stallion, Wilywood. He traded another horse and $200 for the 6-year-old dun and has never regretted it. Not long after he purchased the horse his wife was watching Tom rope on him. She commented, "You'd better keep that one—he's the best you've ever had."

Wilywood earned AQHA points in Reining, Working Cow Horse, Calf Roping, and Team Roping, in addition to carrying Eliason to rodeo wins. The stallion wasn't retired from competition until he was nineteen. He has sired winners in nearly all AQHA events including Barrel Racing, Pole Bending, Roping and the rodeo timed events.

The easy-going stallion also taught all of the Eliason kids to rope.

Several of the Eliason broodmares are daughters of Mr. Flintrock, by Bandit Hancock 1 and out of Flintrock Lady. Mr. Flintrock, with Tom Eliason in the saddle, was 1976 AQHA High Point Junior Calf Roping Stallion. A year later, in 1977, he was AQHA World Champion Senior Dally Team Roping Horse (Heeling) and in the National Top Ten in both Calf

Fig. 10-20. Photo courtesy Tom Eliason.

Tom Eliason of Gregory, South Dakota mounted on **Wily Tuko**, a son of Wilywood out of a Mr. Flintrock daughter. Eliason has been raising Driftwood horses since 1983 when he acquired Wilywood, a double-bred Driftwood stallion.

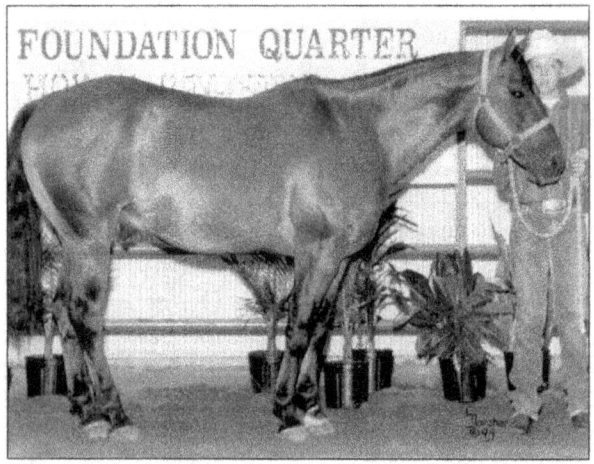

Fig. 10-21. Photo courtesy Tom Eliason.

Wilywood, a son of Orphan Drift out of Oui Oui by Poco Speedy, has been the senior sire at Eliason Quarter Horses since 1983. The good buckskin stallion, originally purchased as a roping mount, has literally put the Eliasons in the Quarter Horse business and made Driftwood believers out of them.

Roping and Working Cow Horse. Crossed with Wilywood, the Mr. Flintrock daughters carry on the Driftwood/Hancock cross that has been so popular with performance horsemen.

While the majority of the Eliason foals go to usin' horse folk, Driftwood breeders are beginning to beat a path to the ranch. A 2000 production sale saw a 1994 Wilywood stallion, Wily Rox, go through the ring at $24,900 to the Fritz Ranch at Niobrara, Nebraska. Under their ownership, Wily Rox is nearing his ROM with points in calf and team roping.

Realizing that Wilywood won't last forever, the Eliasons are bringing on a replacement. This young stallion, Wily Tuko, is by Wilywood and out of a Mr. Flintrock daughter. "Only time will tell us how good a horse he develops into," says Tom. "He'll have to prove that he can perform."

In addition to raising good horses, Tom and JoAnn Eliason have reared a crew of children—five sons and three daughters. "Robert and Beth train outside horses and work with our program," explained Tom. "Both attended college on rodeo scholarships and Blake, the youngest, is in college now on a scholarship."

Summing up their horses is easy for Tom. "Driftwood has been the main emphasis of our breeding program, and it's been awfully good to us."

Babbitt Ranches—Williams, Flagstaff and Ash Fork, Arizona   Located on the high plateau of northwestern Arizona, the Babbitt Ranches have been a part of western history since the 1880s. At the height of their operations, the Babbitts ran thousands of cattle and sheep on 100,000 square miles in three states. Among the brands was the famous "Hashknife," still burned on the left thigh of the horses. They also owned a number of other businesses, from trading posts to an opera house, which buttressed the constantly changing livestock market. Eventually, there were five brothers involved in the various enterprises.

In the 1950s, the ranch operation was consolidated into four holdings: the 635 section CO Bar, the 255 section Cataract, the 225 section Espee, and the Hart (which was later sold).

Today, the Babbitts run approximately 7,800 head of cattle with a remuda of 80 saddle horses, 60 mares and 4 stallions. Over the years, the constant upgrading of the horse stock has not only provided good ranch geldings but a source of added revenue. The horses are strong in the blood of Driftwood.

Prior to 1964, the horses were not registered with the American Quarter Horse Association. At that time, the AQHA opened a hardship clause and the CO Bar took the opportunity to have the mares inspected for registration. Prior to that, only the stallions had been registered.

Bill Howell came to the ranch in the early 1960s and took it upon himself to remember which mares were producing the good horses. "Now," smiles his son, Vic, "We can probably go back four generations. That's what enabled us to do what we're doing today and the reason our horses are so popular. We're raising some pretty good horses and it's because he kept weeding out those bad ones and keeping the tops. Very few ranch people raising horses can say, 'I know the horse's full brother, and he was a good cutting horse, or he was a really good rock-footed horse, or he was extra tough'."

Over the years Babbitt Ranches have used a number of stallions. First, there was Nacho, a registered Thoroughbred purchased from the Greene Cattle Company at Cananea, Mexico. Back in the '60s there was Beaver, who supposedly traced back to

Fig. 10-22. Photo by Jim Morris.

Babbitt Ranches mares and foals. *These horses carry a strong combination of speed, athletic ability and functional conformation which, when crossed with the Driftwood breeding, produces the kind of horses which can cowboy or rodeo.*

Guinea Pig. He was followed by Clabber Boy by Clabber. Other stallions included a grandson of Sugar Bars named Ginger Bars Hank, the Hancock-bred Pocho by Frosty Joe, and Deck Of Wood by Speedy II.

Currently the ranch is using Cowboy Drift by Orphan Drift and out of a Poco Speedy mare; Hanks Chargin Bar; Hanks Star by Tonto Bars Hank out of a King-bred mare; Squeelin To Move by Shot O' Gin and out of a granddaughter of Peppy San; Cowboy Ben Driftin by Cowboy Drift and out of a daughter of Hanks Chargin Bar; and The Double Cowboy by Cowboy Drift and out of a Deck of Wood daughter. The Babbitts also leased Lindsay Peake, by Son of Bar Girl by Son O Sugar and out of a Speedy Peake dam.

Vic Howell is a strong believer in mare power. "It takes top stallions to produce good horses, but the mares are equally important. One thing that you want to be sure to remember about our brood-mares is that everyone has a good brother. If she doesn't produce good colts for us, we get rid of her. And if she's got a bad brother, she doesn't wind up in the mare herd either."

Every prospective broodmare is broken before she goes in the band. She may only get 10 to 15 saddles, but that is enough to tell the ranch if she has the right attitude. This attention to disposition is one of the reasons that the Babbitt horses have found such favor with usin' horse people over the years.

The demand for Babbitt horses is so strong that the ranch has instigated an annual sale each summer. Held at Redlands Camp, north of Williams, the low-key event is well attended by ranchers, foundation breeders, ropers and horse people. The 30 colts and fillies are separated from their mothers, numbered and branded with the famous Hashknife. Then, with Vic Howell and Babbitt Ranches CEO Billy Cordasco spotting bids from the center of the corral, they are auctioned off. The foals are then reunited with their mothers and turned out to pasture until the following March, when they will be weaned and picked up by their new owners.

Babbitt horses are characterized as having good dispositions and lots of cow, and being smooth travelers which can really get across the country. They cinch big, have good withers, straight legs,

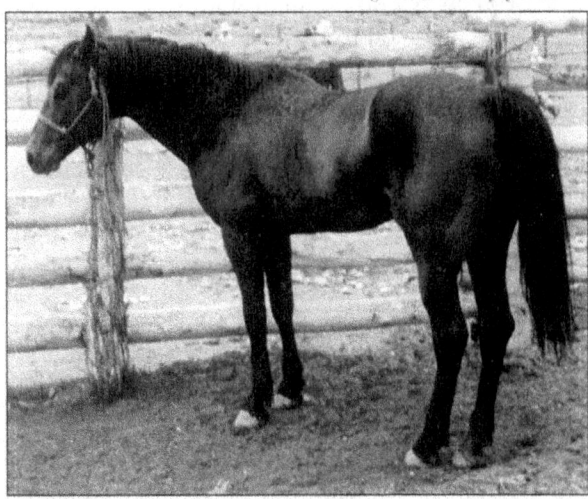

Fig. 10-23. Photo by Jim Morris.

Cowboy Drift, *one of the Babbitt Ranches stallions, in range shape. He is by Orphan Drift and out of Poco Judy Sue out of Judy Sue by Driftwood.*

plenty of bone, and nice heads with plenty of room between the eyes.

"We're after a rodeo type horse that will work on a ranch," comments Vic Howell. And judging by the demand for the Babbitt horses, they have what the public wants.

Arnold Quarter Horses—Hutchinson, Kansas Like many Driftwood breeders, the Arnold family became associated with the horses through the rodeo arena. In the 1970s Perry Arnold was rodeoing and roping. He saw many contestants from California and Arizona riding Driftwoods. Julio Moreno's Six Pac, Jim Rodriquez Jr.'s Gill, and Brad Smith's mount were all the kind of rodeo horse he admired. He'd read about the Driftwoods, and heard other ropers talk, but until he saw them in action he didn't realize just how talented they were. He knew then that if he ever raised horses it would be the Driftwoods.

In 1988 that opportunity came. Because of a football accident to his son J.R., the Arnolds moved to Hutchinson, Kansas and purchased a 640-acre ranch. Since it was not big enough to carry a sufficient number of cattle, the family decided to concentrate on raising the best performance horses possible—and that meant Driftwoods.

Perry contacted old rodeo friends, asking where he might find horses of Driftwood bloodlines. After

Fig. 10-24. Photo courtesy Arnold Quarter Horses.

Horses are a family affair with the Arnolds. **Zane Arnold** *aboard the ranch stallion* **Lucky Goldwood** *(Wilywood/Clark's Doc Bar daughter),* **Dusty Arnold** *and* **Perry Arnold** *in the background.*

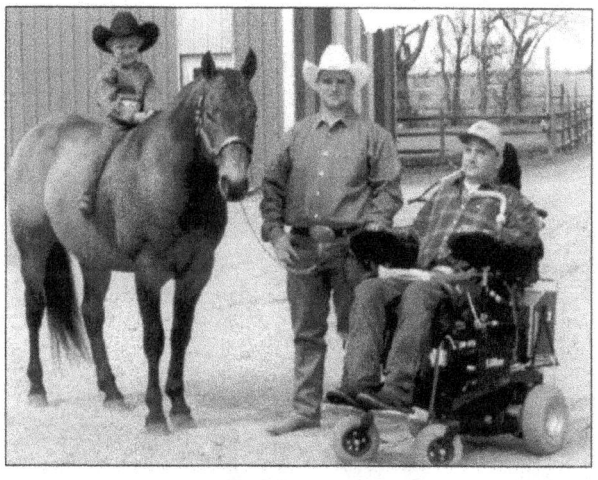

Fig. 10-25. Photo courtesy Arnold Quarter Horses.

Zane, Dusty *and* **J.R. Arnold** *with* **MJH Rosewood Ike** *(Wayward Ike by Driftwood Ike out of a Cibecue Roan daughter).*

Fig. 10-26. Photo courtesy Arnold Quarter Horses.

The roping arena is where the Arnold story started and the family is still at it. **Dusty Arnold** *heading on a Driftwood-bred gelding and father* **Perry** *heeling on a palomino son of Wilywood.*

lots of telephone time and criss-crossing California, Arizona, Utah, Wyoming, and the Dakotas to visit breeders, he and Martha obtained some Driftwood mares. The majority of those individuals not only possessed the breeding and conformation which the Arnolds felt were desirable, but "could do something under saddle" as well. Those mares were crossed with stallions already owned by the Arnolds. A two-year-old double Driftwood-bred stallion was also purchased and added to the stud battery. The resulting foals were what they were looking for, and what other ranchers and ropers wanted. Before long, another Driftwood-bred stallion was acquired.

At the present time, the operation has between 45 and 50 mares of Driftwood, Joe Hancock, Lucky Blanton, and Joe Reed II breeding. Early in the program the family decided not to breed outside mares, so all the foals are strictly a product of the Arnold program. "At the present, we're selling the colts right off the mares," remarks Perry Arnold. "People are calling and reserving colts out of a specific mare, even before the foal is born. Our entire 2000 foal crop was gone before the end of the year."

Four stallions, MJH Rosewood Ike, CC Blue Driftwood, Lucky Goldwood, and P C Willyboy, ranging from 12 to 28 percent Driftwood, are used by the ranch. They are also riding four young stallions of Driftwood breeding, testing their ability under saddle before turning them with the mares. All the stallions run out with the mares when they are not being used under saddle. Unlike many breeders, the Arnolds "haul" their studs, rodeoing and roping on them. "When we catch them out of the pasture, they know that they're going to be ridden," smiles Perry Arnold. "We can tie them up next to a gelding at a roping and they'll never make a sound."

In addition to producing horses with speed, ability, conformation and disposition, they are adding color. Blue and red roans, buckskins, palominos, blacks, grays, and duns predominate. "We didn't start out with that in mind, but the genetics of the horses we liked seemed to produce color," commented Arnold. "We're trying to raise the best type of horse possible and keep as much Driftwood in the pedigrees as we can. The color is an added bonus."

Arnold Quarter Horses is a family operation with Perry and Martha Arnold running the business, searching for more Driftwoods and handling client contact. Son J. R. takes care of bookkeeping, photography and advertising. Another son, Dusty, and his wife, Stephanie, along with daughter Tamra, train and promote the horses by attending ropings, rodeos and barrel races.

In raising the Driftwoods, the Arnolds are producing horses that follow the old roper's belief: "You gotta be mounted to win."

Mike and Arlene Pixley—4P Cattle Company, Upton, Wyoming Mike Pixley has strayed away from horses a few times in his life but always came back. He was raised on the Ox Yoke Ranch at Nemo, South Dakota, and after finishing his education, drifted away to another occupation. Then, in 1977, he and wife Arlene moved to Sheridan, Wyoming. He took up chariot racing, a sport not noted for those short on nerve. For seven years he raised, trained and raced chariot horses, not just for himself but for other people as well. His children were also involved in 4H and junior rodeos and the equine commitment got pretty deep.

The family moved to Gillett, Wyoming in 1991, bought a ranch in '92, and bred their first mare in 1994. At first the Pixleys ran about 250 head of commercial cattle with the horses as a sideline. It wasn't long before the demand for their horses crowded out the cattle. Currently the 4P Cattle Company maintains approximately 120 mares and colts which they keep back for future rope horses or replacement broodmares. Some 70 foals are marketed every year during the annual sale, along with thirty finished roping geldings. The program has been strongly influenced by Driftwood, Lucky Blanton, and Joe Hancock. Other bloodlines represented are Oklahoma Fuel, Peppy San Badger, and J B King. Pixley says, "Most of our mares carry a high percentage of Foundation blood. We have granddaughters of Bert and King and nine own daughters of Orphan Drift. I think that the Driftwoods are some of the greatest horses I have ever known."

Those mares are run outside, pasture-bred and pasture-foaled. That, coupled with the strong genetic base, gives them the kind of background to raise useful horses. The foals grow up outside, with plenty of room to run and develop.

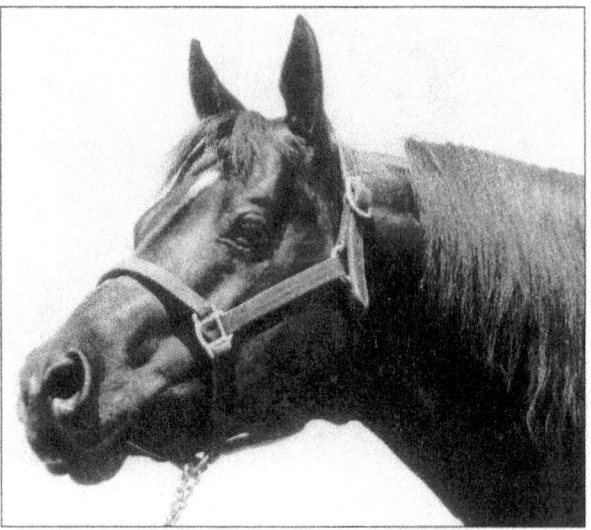

Fig. 10-27. Photo courtesy 4P Cattle Company.

Fast Commander, *one of the stallions at the 4P Cattle Company, carries a generous dose of Driftwood in his pedigree. As Mike Pixley says, "I think the Driftwoods are some of the greatest horses I've ever known."*

Fig. 10-28. Photo courtesy 4P Cattle Company.

Rojo Red Drift, *by Juana Rojo, is another 4P stallion strong in Driftwood blood. The roan stallion is a seasoned rodeo mount and carries Jay Pixley to lots of calf roping money.*

The first stallion purchased was Barlons Grandad, a grandson of Old Grandad. He was followed by a son of Wilywood and then Fast Commander. The latter was followed by Lucky Hug Tonto, a grandson of Lucky Blanton; and then by Rojo Red Drift, a grandson of Blue Valentine out of a Captain Crusade/Candywood mare. The newest stallion is Snicklefritz Chex, a grandson of King Fritz and out of an Eddie 80 daughter. Currently, the Pixleys are standing a battery of twelve stallions featuring a combination of performance bloodlines.

All the stallions are well-broken, and with the exception of the older ones, ridden regularly for ranch work and roping. The 4P Ranch believes that not all young stallions are "stud material" and are quick to geld one that doesn't learn willingly or have the athletic ability to be a top using horse. All young horses are ridden as two-year-olds to determine if they will fit in the breeding program. That doesn't mean just arena riding—they go outside for ranch chores and see lots of cattle.

The 4P Cattle Company held its first production sale in 1999. Buyers from 18 states were on hand to bid on the offering of 85 4P horses and 26 from Mills Quarter Horses of Sundance, Wyoming. That was followed by another successful sale in 2000 with a number of repeat buyers. "We try to offer 30 to 35 finished rope horses, mostly team roping mounts but a few calf horses, each year," comments Mike Pixley. "The older geldings carry a satisfaction-or-return guarantee, and we haven't taken one back yet."

Absolute honesty about every horse offered for sale is the watchword at a 4P sale. "I'm very conscientious about putting down everything about a horse in the catalog," Mike says. That fact, coupled with the quality of the offerings, insures repeat customers.

Family involvement is another factor at the 4P. In addition to Mike and Arlene being on hand all the time, daughter Stacey Blakeman of Cheyenne comes home to help prep horses for the sale. Jay is attending college in Texas and competes on Rojo Red Drift in calf roping. He spends his summers on the ranch, working with the horses and getting ready for the sale.

The Pixley's 4P Cattle Company is dedicated to raising and developing horses which typify the Quarter Horse breed and will be the "legends" of the future.

John Rudnick—Santa Margarita, California The old saying "Scratch a cowman and you'll find a horseman" has never been more true than with this native Californian. John Rudnick grew up on the family ranch and was exposed to good horses from the time he could ride. He picked up a rope as a young boy and has never recovered from the itch to compete. Long-time friend Tex Oliver introduced young John to the Driftwoods in 1984. Ever since then, John has used the horses extensively on the ranch to work cattle. When they go to the roping arena they are prepared for everything.

In addition to extensive cattle ranching operations, John Rudnick has established a Quarter Horse operation dedicated to producing rodeo rope horses—just as Katy Peake did over half a century ago. The foundation sire of the program was Frostys Tops, by Pelican, a top rope horse himself and a sire of rope horses with earnings of over $2.5 million at the professional level. His foundation mare was OH April by Driftwood Ike, whose get have roping earnings well over $200,000.

All the John Rudnick mares are strong in the blood of Driftwood, War Concho and Frostys Tops. The three stallions standing at the Rudnick ranch mirror this breeding. SG Rodeo Cowboy is by Frostys Tops out of OH April by Driftwood Ike; JR Tigerwood is by Dynaflow Drift out of OH April by Driftwood Ike; and Amarillo Drift is by Flipwood out of Doc A Ling by Dry Doc.

Jerold and Leo Camarillo, World Champion Team Roping legends, are both riding full brothers out of

Fig. 10-29. Photo courtesy John Rudnick.

SG Rodeo Cowboy, Frostys Tops by Pelican out of OH April by Driftwood Ike, is the senior stallion at Rudnick Quarter Horses and is siring some outstanding usin' horses.

John Rudnick's favorite mare OH April. One of these individuals, Double Tough San (Sakum San/OH April by Driftwood Ike), owned by Jerold and ridden by Leo, is a NFR-caliber mount, capable of heading or heeling and has roping winnings in excess of $100,000. The other Sackum San/OH April gelding owned by Jerold is Lariat, another pro-level rope horse.

OH April, when bred to Call Me Ike, produced Casper for Sherrick Grantham who sold the horse to Clay O'Brian Cooper, 7-time World Champion Team Roper. Bred to Call Me Ike and in foal with Ikes April, OH April was sold to John Rudnick in 1994. Since then the prolific producer has foaled Ikes Frosty Girl and SG Rodeo Cowboy, both by Frostys Tops, John's number one stallion. Another Frostys Tops mare out of Annie M. Brown by True Blue Brown, owned by Rudnick, is a head horse all the way, with winnings in excess of $45,000.

John Rudnick is also aware that it takes good cowboys to bring out the potential in a horse. He has two of the best—Leo and Jerold Camarillo, World Champion Team Roping legends—hauling his geldings down the road. One of these individuals, Double Tough San (OH April by Driftwood Ike), is capable of heading or heeling and has roping winnings in excess of $100,000.

In raising Driftwoods and aiming at producing the kind of horses that ropers can win on, John Rudnick is following a well-established tradition.

Henning & Suzanne Koch—Stevinson, California
Over the years a soft-spoken cowboy and roper and a horse-crazy girl from Salinas have developed a successful horse operation combining the proven performance of the Driftwoods with the color of Appaloosas. Henning Koch was a money-winning calf roper who knew the kind of horse he wanted and always rode them. His wife Suzanne grew up in Salinas, learned to ride with Marvin Roberts Sr., and embarked on a successful show career.

Suzanne's first show horse was a chunky white mare, Cricket, which she showed in stock horse classes. After the girl graduated from high school she bred the mare to a bay Quarter Horse stallion. The foal was an Appaloosa. Cricket went on to produce twelve foals, all by Quarter Horse or Thoroughbred stallions and all of Appaloosa color.

Fig. 10-30. Photo courtesy John Rudnick.

Usin' them outside at the Rudnick Ranch helps make good horses. **Driftwood Magic**, SG Rodeo Cowboy/Christa Three, makes a hand trailing cattle.

Fig. 10-31.

Henning & Suzanne Koch and **Double or Nothin'** with High Point Saddle of Mid Valley Show, 1978.

Fig. 10-32.

Right: **Henning Koch on Kittywood** and left: **J. D. Yates on Birthday Girl** after winning the team roping at the 40th Annual National Appaloosa Show in Albuquerque, New Mexico.

One of those foals was the Appaloosa stallion Double or Nothin' by Sugar Bars. He was the first Triple-A rated Appaloosa race horse, then did everything a performance horse should and was finally a Nationally Rated Appaloosa Sire of Performance Horses and in the National Hall of Fame.

Growing up in central California exposed Suzanne to the Driftwoods, and she always knew they were the type of horse she wanted to own and show. As a roper, Henning was already aware of them since so many of the men he competed against rode Driftwood horses. It took some time, but they eventually were able to purchase one, Tiny Tot Peake. That Driftwood daughter, crossed with Double or Nothin', gave the Kochs the kind of horses they wanted—with the dash of color.

Over the years Henning and Suzanne Koch's Double or Nothin'/Driftwood or Double or Nothin'/Speedy Peake bred horses have successfully competed in Cutting, Reining, Barrel Racing, Roping, Racing or any other event in which their owners have been interested. And they have won at Open, National and Professional levels.

A daughter, Sherri, lives in Texas and competes at Appaloosa shows as well as Professional Women's Rodeo Association contests. One of her main competition mounts is an Appaloosa mare named Sugar Peake.

Building on the framework of a well-bred Appaloosa stallion and a good-producing Driftwood daughter, Henning and Suzanna Koch have built up a breeding operation that stresses quality, performance and color.

Some of these individuals are line-breeding, keeping the Driftwood line as pure as possible. Others are utilizing the known performance potential and out-crossing, breeding their Driftwood horses to other families within the American Quarter Horse breed. Regardless of the manner in which Driftwood blood is infused into a pedigree, the traits which have characterized the family for over half a century continue to predominate.

11

A NEW CENTURY

"Driftwood was one of the greatest horses I ever knew. How many stallions have stood the test of time as he has—and the current revival of Driftwood interest in carrying on something which we knew fifty years ago."

—Perry Cotton

By the 1990s, popular interest in the Driftwoods surfaced again. Quarter Horse producers were returning to the foundation breeding which had made the breed and were looking to the breeders who had maintained the time-proven performance bloodlines.

A few breeders, primarily in California, Arizona, Idaho and the Dakotas, had hung onto the Driftwood line. The foals which they produced had gone to other ranchers and ropers with little fanfare. Then, horsemen began to return to the source, looking for the kind of mount which could perform, had a willing mind, and would stay sound. Since a number of the people coming into the business had been (or still were) ropers who knew of the Driftwood reputation, it was natural that they would contact their friends for breeding stock. The Easy Keepers, the Driftwood Ikes, the Speedy IIs, the Speedywoods and the Orphan Drifts had kept the strain alive and the cowboys were looking for them. As more and more individuals took the horses out before the public the Driftwood story began to spread out to people who had never heard of the bay stallion.

1994 Western Horseman book *Legends II*, in which Driftwood was profiled along with other early stallions, gave added emphasis to the rise of interest. Driftwood again became a well-known name to the general horse public. Horsemen began searching for the scattered breeders who had maintained the line and infusing Driftwood blood into their programs.

Gene Moench of Valpairaiso, Indiana wanted some new blood (or old blood) in his performance horse program. He had learned of the Driftwoods and wanted to know more. His first step was to visit Jim West of Ione, Oregon and see Drifts Chip. That started his journey. Since Arizona, California and the Dakotas seemed to be the main source, Moench began to make inquiries about who was also breeding the horses. He visited

Fig. 11-1. Photo by Livingston.

Driftwood fans *from all over the West gather at the Will Gill & Sons Sale, October 20, 2001, Madera, California.*

with Henry Kibler in Arizona, Jim Morris and Will Gill, Jr. in California, and any other breeders he could locate. He found that there were more Driftwood-bred horses out in the country than he realized. As he located the type of mares he was searching for, he added them to his program.

Through established breeders of the Driftwoods, Moench learned of others who still had them. He made phone calls, visited old-time horsemen and women and looked at lots of horses. If a little knowledge was good, a lot was better. He began to feel that the knowledge and history of the strain should be shared.

In 1994, the enthusiastic Moench made arrangements to meet with Jim Morris, Henry Kibler, Will Gill, Benny Norman, Sammy Fancher Brackenberry, and Shannon Price at the 1995 Snaffle Bit Futurity in Fresno, California. The gathering was to discuss the formation of a Driftwood Breeders Association. He also teamed up with Jim Morris and Brian Wright to see just how much Driftwood blood there was still around, and how many people were breeding the horses. Those three individuals felt that the project was worth pursuing and placed an advertisement requesting information on the Driftwoods in the *Quarter Horse Journal*. That ad received an enthusiastic response. Letters poured in from around the country with comments and praise for the horses. Some of these are quoted:

"My experience shows them to be very cowy, very smart, willing to work...but need to be treated with respect." —Texas

"In all the years I've had Quarter Horses, I've always had Driftwoods. They are wonderful. I still have one." —Oregon

"Just thought I'd drop a note as we are certainly high on the Driftwoods. Their working ability is unreal! Just wish we had more of them." —South Dakota

"I owned a grandson of Driftwood. We used him for everything on the ranch, roping, cutting, and also barrels, poles, bulldogging and reining. He could do it." —Texas

"You're right! Driftwoods had more sense than most of their riders." —Arizona

"I've been keeping files on the foundation horses for over 25 years and have some of what I believe you're looking for. These are articles in the *Quarter Horse Journal* on Driftwood, Sage Hen, Poker Chip Peake, Henny Penny Peake. Hope this has been some help. If there is anything else, let me know." —Wisconsin

With responses like that to encourage them, Moench, Morris and Wright decided to formulate and send out a questionnaire to get more information.

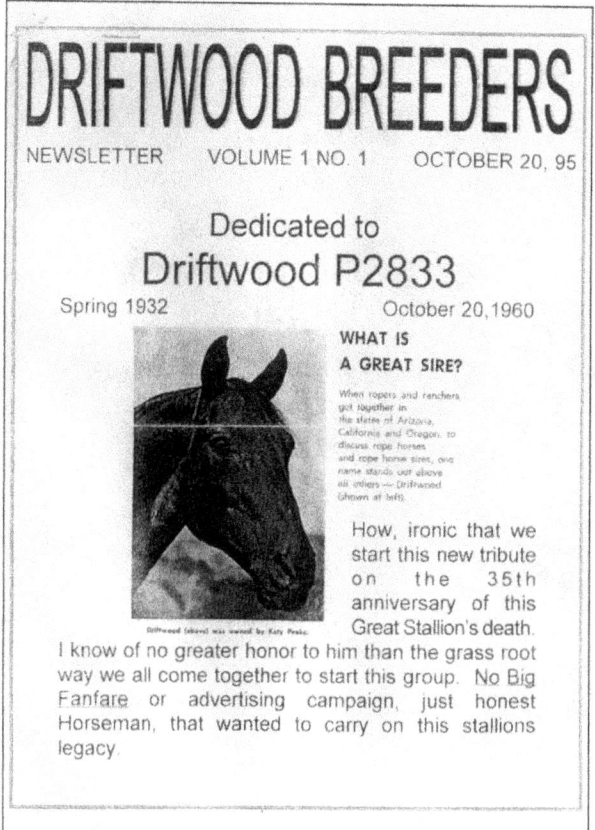

Fig. 11-2. Courtesy Jim Morris.

*Volume 1, Number 1 of the **Driftwood Breeders Newsletter**. It was dated October 20, 1995—exactly 35 years after Driftwood's death. Edited and produced by Jim Morris, this newsletter had a positive impact on bringing Driftwood breeders together and adding emphasis to the swell of interest in Driftwood's descendants.*

That questionnaiare asked how the accumulated knowledge should be shared with other horsemen, what the recipient's Driftwood experience had been, what crosses with other bloodlines were compatible, how they used their horses, what was liked about the horses, and also asked if there were stories they wished to share. Over 80% of those receiving the questionnaire responded and the ball began to roll. All questionnaire recipients were invited to a "get together" at the 1995 Snaffle Bit Futurity in Fresno, California.

Those attending the October 21, 1995 meeting included Gene Moench, Brian Wright, Jim Morris, Will Gill Jr., David Gill, Bennie Norman, Steve Branco, Henry and Nancy Kibler, John Rudnick, Russell Keeley, Shannon Pearce, Tyler Morris, and Michle LeClerc. Perry Cotton and Sammy Fancher Brackenberry were unable to attend because of conflicts. Other prominent horse people who expressed their support but were unable to attend were Jim West, Chuck and Shiela Biggerstaff, and Jim and Emily Averill.

That gathering of Driftwood fans resulted in the informal Driftwood Breeders Association. Jim Morris assumed the responsibility of producing a quarterly newsletter which would be funded by breeder advertising. The publication contained memories of Driftwood owners, stories and photographs of Driftwood get, and biographies of various breeders, as well as a listing of current ranches producing Driftwood horses. At last the legend was beginning to take unofficial shape. The newsletter was followed by a *Directory of Driftwood Breeders and the History of Driftwood*, compiled and published by Brian Wright. Not only was the history of the Driftwood line becoming public knowledge, but the accomplishments of those who had carried it were recognized. At last, horsemen learned of other Driftwood breeders and were able to network.

Old stories about Driftwood and his get, forgotten photographs, and remembrances of the men and women who had bred and used the horses all began to surface and were passed on in print. The inclusion of the newsletter in the *Foundation Quarter Horse Journal* and recognition of the Driftwood contribution in the *Quarter Horse Journal* and other equine publications all contributed to the rising popularity. What started as a swell became a wave as more and more people discovered the qualities of the Driftwoods. No longer are they an open secret among ropers and ranchers. Prices on ranch and rodeo horses carrying Driftwood bloodlines soared, such as the 1999 Cowan sale with Lone Drifter colts averaging $14,000 and the 3-year-old stallion, P C Frenchman, bringing $200,000.

Established Driftwood breeders, those who raised the horses in volume, scheduled sales highlighting the blood. A number of these were to become annual affairs, well-attended by interested horsemen who paid premium prices. Sales such as Eliason's, Pixley's, Cowan's, Will Gill and Sons, West's, and Arnold's all brought horsemen together from all over America. In each case, the affairs took on a quasi-social touch as buyers and spectators gathered to "talk Driftwood." Individuals who had just heard of each other became friends through the common denominator of the horses they loved.

Fig. 11-3. Courtesy J. N. Swanson.

'Branding—High Desert' by J. N. Swanson. This California vaquero has painted the Driftwoods for fifty years in addition to raising, training and exhibiting stock horses. Note the "Driftwood look" on the lefthand horse.

As they have for more than a half-century, the Driftwoods continue to proliferate in the rodeo arena. Ropers, steer wrestlers and barrel racers continue to look for the proven performance bloodlines, functional conformation, quick speed and willing minds that can take them to the pay window. Professional rodeo, the National Finals Rodeo, the big team ropings and barrel races are all showcases of the bloodline, although usually little is said of how a particular mount is bred. To a Driftwood fan, one look is usually enough.

That "Driftwood look" has also appealed to another venue. Artist Jack Swanson of Carmel Valley, California is a long-time fan of the horses and depicted them in many of his paintings. As a vaquero working on such well-known California ranches as the Crofton Cattle Company, as a successful bridle horse trainer and long-time breeder of good stock horses, Swanson has had an ample opportunity to discover the special conformation, the athletic ability and try that typified the Driftwoods. "They're the right kind of horses, the kind that everyone likes to ride. I like to paint them," comments the artist. Swanson's work depicts the ranching scene of the Pacific Coast, both today and back through history. And his horses all have the Driftwood appearance and carry the trappings of the California vaquero.

Present day breeders have taken up the challenge, and with the strong bloodlines still available, are continuing to produce one of the greatest lines of performance Quarter Horses which ever existed. Driftwood fans are also taking advantage of unusual events to showcase their horses, exposing them to people who have never had the opportunity to see them before.

During the preliminary activities for the 1996 Olympic Games a cross-country torch relay ride was staged. The Olympic torch was carried, via runner, bicycle rider and horseman, from Los Angeles, California to Atlanta, Georgia—the site of the summer games. In cooperation with the Atlanta Olympic Organizing Committee and the National Pony Express Association, three Driftwood horses were trailered to Kansas to gallop a segment of the original Pony Express Trail. Wilywood, owned by Tom Eliason; Colonel Charge, owned by Clair Jones; and Chips Gitano, owned by Brian Wright, all showed thousands of spectators just how classy the Driftwoods are. The blazing torch above their heads, cheering crowds, and unfamiliar riders were handled with typical Driftwood aplomb. As one spectator commented, "Those were probably the finest horses ever to travel that part of the original Pony Express Trail."

Fig. 11-4. Courtesy J. N. Swanson.

'Culling the Herd' by J. N. Swanson. Another Driftwood horse doing what they do best—working cattle. Artist Swanson can not only paint them but use them as well, since all of his horses are trained in the "Old California" style.

In 1997 the Driftwood Breeders Association proposed the installation of a Historical Marker honoring Driftwood to the American Quarter Horse Association Historical Markers Program. Such markers honor places, people, and horses which have been significantly prominent in the development of the Quarter Horse. The plaque would be located on a highway next to the Rancho Jabali property and honor both Driftwood and Katy Peake for their contributions to the performing Quarter Horse in the Western United States. While the current owner of the property was agreeable, the AQHA did not act on the request.

In 2001 two of California's leading horsemen were honored by induction into the National Reined Cow Horse Association Hall of Fame. Both men, legends in Pacific Coast show arenas, were mounted on Driftwood-bred horses during their careers:

Ronnie Richards of Irvine, California is credited with developing more international and American Horse Show Association Medal Champions than any other horseman. His winner list reads like the "Who's Who" of the stock horse world. Among those great individuals was Jimmywood, by Super Charge and out of the spectacular Henny Penny Peake. On him, Richards won the 1966 Santa Barbara (California) Nationals Championship. In 1977 he took the Cow Palace Mare Class on Speedy Cash, by Cibecue Roan who was by Red Man and out of Miss Linwood by Driftwood.

Jimmy Williams of Flintridge, California is considered one of the greatest horsemen to ever step across either a stock horse or a hunter/jumper. Among the great hackamore and reined stock horses he trained and exhibited were Henny Penny Peake and Woodwind (Hot Toddy), both by Driftwood. His comments on the Driftwood ability are in the opening chapter of this book. In 1960, Williams was named "Horseman of the Year" by the American Horse Show Association.

He has been recognized by three California Governors for his contributions to the sport of horsemanship. In the 1960s, Williams changed his focus to hunters and jumpers, but they could all stop and turn around.

As both Ronnie Richards and Jimmy Williams well knew, "You have to be mounted to win." With the Driftwoods they were.

Without a doubt, the most important monument to the bay stallion and to the dedicated woman who owned him is the continued popularity of the bloodline. Katy Peake strongly believed in Driftwood, and to her death continued to feel that he had made a lasting contribution to the performance horses of the West. Each time a roper backs a Driftwood into the box, each time a stock horse contestant takes a cow down the fence with a Driftwood or a barrel racer points her Driftwood at the course, they are carrying on the tradition which began in 1943 that day in Phoenix when Katy Peake loaded the bay stallion into a trailer.

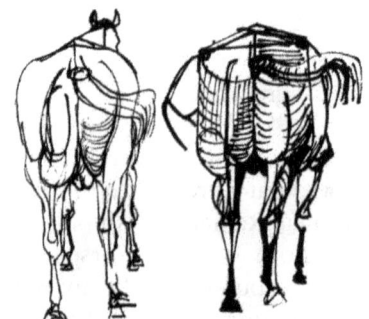

APPENDIX 1

AQHA PROGENY RECORD
DRIFTWOOD

AQHA Registered Foals	152
No. of Foal Crops	25
Performance Foals	31
Crops Raced	3
Number of Starters	2
Number of Starts	5
Register of Merit	1
Leading Money Winner	Calzona Katy No. 55,947
Points Earned in All Show Divisions	278.5
Point Earners in All Divisions	29
Halter Points	34
Halter Point Earners	29
Performance Points	242.5
Performance Point Earners	5
Registers of Merit	19

Since the majority of Driftwood foals which were raced ran as match horses there is no record of results or winnings. A large percentage of his offspring went to the rodeo arena where their winnings are not recorded or were shown at events sanctioned by the American Horse Show Association or the California Reined Stock Horse Association and the results are not tabulated by the American Quarter Horse Association.

Breeder:
Mr. Bailey Childress—Silverton, Texas

Owners:
Sam Turner—Silverton, Texas
Amos Turner—Silverton, Texas
Buck Nichols—Gilbert, Arizona
George Cline—Payson, Arizona
Asbury Schell—Tempe, Arizona
Catherine Peake—Lompoc, California

APPENDIX 2

AQHA GET OF SIRE LISTING
DRIFTWOOD

The following lists the 152 horses sired by Driftwood and registered with the American Quarter Horse Association. Undoubtably, there are a number of foals sired by him before the formation of the AQHA in 1941 which were never registered, so this is not a completely accurate list of his offspring. The very fact that Cowboy Schell was foaled in 1935, when Driftwood was just a three-year-old, Clayton Gal and Smokey in 1936 when he was four, and Goober and Snake when he was five indicates that the Turner family was breeding the stallion at an early age.

Number	Name	Foaled	Sex	Dam No. & Name	Number	Name	Foaled	Sex	Dam No. & Name
AO43421	Speedy Bar		S	32032 Rhoda Bar	0016612	Cowboy Schell 1943 - Racing ROM	35	G	U075288 Will Steed
NO03223	Dixie Bell Natl		M	QO51108 Dixie Bell	0016695	Woodrust	47	G	0014560 Coal Black
UO74714	Mare by Driftwood		M		0017045	Speedy N	45	G	0002192 Nosey
UO79193	Sally Wood		M		0017278	Katy Red	46	M	0002193 Prairie Dog
U231335	horse by Driftwood		S		0017337	Fleet Wood	46	M	0002189 High Tone
0003250	Wanda Marie	39	M	Mare by Miller Boy	0018660	O See O	46	M	U074004 Little Moore
0003367	Clayton Gal	36	M	UO78492 Red	0019955	Wood Wasp	46	M	0002196 Waspy
0005299	Melody Boot	41	M	U14681 Dolly	0020404	Jima Red	45	M	0002191 Jack Rabbit
0006629	Dogwood	44	S	0002518 Smoky McCue	0021702	Baywood	48	M	0014560 Coal Black
0006668	Drifty	43	S	0002194 Sage Hen	0022400	Driftalong Katy	50	M	0004884 Freya
0006751	Mac McCue W 1957 - Open ROM	43	S	0002518 Smoky McCue	0022401	Northwood	50	S	0004886 Hildegund
0007059	Tres X	44	M	0002517 Dos X	0022531	Buckwood	48	M	0005593 Hancock Belle
0007366	Cherokee Jake	44	S	U078001 Pigeon	0022532	Driftwood II 1957 - Open ROM	50	G	0005619 Shu Cat
0008062	Little Speed	44	S	0002192 Nosey	0022534	Henny Penny Peake PCHJSHA and AHSA Stock Horse of the Year	49	M	0002194 Sage Hen
0008076	Sky Boy H	45	S	0002328 Skylark	0022536	Poker Chip Peake 1957 - Open ROM	50	S	0002194 Sage Hen
0008509	Cowboy C	40	S	U299307 Mickey					
0008848	Roadrunner P	44	M	0002194 Sage Hen	0023197	Two Pence	49	M	0002523 Sister Penny
0009174	Stinger B	45	M	0002190 Hornet	0024647	Tonita Hall 1957 - Open ROM	45	M	0008073 Tonita H
0009363	Nellie S	39	M	U156310 Trixie					
0009451	Clara Bow M	38	M	U156310 Trixie	0024752	Allen's Miss Skyla	47	M	0002328 Skylark
0009453	Rosy D	42	M	U156310 Trixie	0025496	Sequoia Wood	50	M	0022404 Gravy
0009455	Flickers S	43	M	U156310 Trixie	0026631	Bill Adair	49	G	0005579 Mayflower
0010757	Blue Glory	45	M	0002195 Straw	0026909	Wood Tick	49	S	0009398 Miss Hondo
0010758	California Sweetheart	45	M	0002517 Dos X	0027477	Miss Drifty	49	M	U074200 Louise
0010781	Red Button	45	M	0002194 Sage Hen	0027977	Mockey Bay	43	M	U081303 Walls TB mare
00119	Jabalina P	43	M	0002517 Dos X	0028563	Coal Bin	49	G	0014560 Coal Black
0012044	Easy Keeper	45	S	0002518 Smoky McCue	0029870	Slick Brown	44	G	U081302 Walls mare
0013401	Dusty Boy	46	G	0002327 Dusty Miller	0031149	Greasewood	51	M	0004886 Hildegund
0013402	Brownwood	46	G	0001321 Brown Baby	0031475	Smoky Cue	50	G	0002518 Smoky McCue
0016570	Scooter Mac	45	G	0002193 Prairie Dog					

Number	Name	Foaled	Sex	Dam No. & Name	Number	Name	Foaled	Sex	Dam No. & Name
0031504	Keetawood	50	M	0008271 Keeta	0056821	Hug Me Tight	55	M	0008456 Nugget Hug
0032525	Miss Linwood	51	M	0002781 Queen Ann	0056971	Speedywood	56	S	0005593 Hancock Belle
0032960	Sierra Jetsan	49	M	A041449 Sierra Buckskin Mary		1966 - Open ROM			
					0057009	Calzona Babe	55	M	0011789 Smarty Pants
0034066	Chilina	50	M	0005593 Hancock Belle	0057828	Bobbinet Peake	54	M	0028869 Bobbie Pin
0036169	Smoke Maker	51	G	0002518 Smoky McCue	0058912	Filet Mignon	55	M	0015901 Jamboree
0036318	Swiftwood	52	M	0013359 Spiderette	0062027	Oak Wood	56	M	0026719 Oklahoma Gal
0038362	Firewood	52	S	0008271 Keeta 7	0062167	Speedy Peake	56	S	0005619 Shu Cat
	1957 - Open ROM					1961 - Open ROM			
0038833	Brown Beulah	52	M	0002781 Queen Ann	0062168	Tweedle Dee	56	M	0052311 Tuckaluck
0041352	Gadget	50	M	0040458 Louise Schuyler	0062915	Star Shooter	56	G	0011742 Shooting Star
0041510	Woodkee	53	M	0008271 Keeta 7	0063070	Bugaboo Peake	55	M	0015900 Honey Bug
	1961 - Open ROM				0063265	Quick A Lick	56	M	0008456 Nugget Hug
0041535	Dusky Peake	53	S	0034067 Dusky Ruth	0063470	Kitty Wood	55	M	0005619 Shu Cat
0042166	Peakewood	53	G	0005619 Shu Cat	0063577	Annie Wood	56	M	0002781 Queen Ann
0042198	Driftalena	53	M	0014853 Lady Lux N	0063993	Wood Mite	56	M	0013359 Spiderette
0042773	Nifty Wood	52	G	0014853 Lady Lux N	0064272	Misty Miss	55	M	0042290 Reed's Hannah Hancock
0043294	Drift Easy	53	S	0002518 Smoky McCue					
0043632	Miss Wood	52	M	0004352 Miss Limit	0064503	Armell	52	M	0009399 Sugar Tit
0043872	Nodoze	51	G	0009399 Sugar Tit	0065035	Speedy II	55	S	0034067 Dusky Ruth
0044426	Drifting Sage	54	S	0002194 Sage Hen	0066014	Speedy Darnell	56	S	0034067 Dusky Ruth
0044578	Quita Bonita	54	M	0001805 Sweetwater Sue	0066244	Rose Wood	54	M	0009398 Miss Hondo
0045052	Woody	52	G	0010772 Betsy Luck	0069686	W P R Lucky Drift	50	M	0015157 Miss It
0046864	Driftwasp	54	S	0002196 Waspy		1965 - Open ROM			
0047174	Jay Wood	50	G	0008091 Black Maria D	0072927	Curleywood	57	S	0008687 Belle Blake
	1960 - Open ROM				0074813	Zoe Driftwood	57	M	0002517 Dos X
0047617	My John	53	G	0010772 Betsy Luck	0076442	Sierra Drifter	57	G	0042033 Sierra Sue Joe
0047645	Driftwood Ike	54	S	0005593 Hancock Belle	0076443	Sierra Redwood	57	M	0043302 Sierra Roxie
	1960 - Open ROM				0076444	Sierra Speedy	57	G	0003541 Tonia V
	1969 - 2nd High Point Steer Roping				0077753	Tooka Peake	56	M	0025720 Lucky Pat
0047646	Woodfern	54	M	0008271 Keeta 7	0078206	Wood Mist	57	S	0034067 Dusky Ruth
0048005	Chakaty	54	M	0014853 Lady Lux N	0078235	Drift Girl	50	M	0005310 DR Charlotte
0048030	Jernigan Peake	54	S	0013359 Spiderette	0079726	Speedy Hug	57	M	0003876 Wonda C
	1964 - Open ROM				0080075	Peake	55	G	T061222 Missmark TB
0048556	Cotton Cat	54	M	A012685 Katy King		1963 - Open ROM			
0050017	Wood Start	49	M	U067956 Beater	0080515	Judy Sue	55	M	0008092 Nita H
0050593	Speedy Cat	54	M	0005619 Shu Cat	0081493	Tweedle Dum	56	M	0034821 Lucky Lady Tucker
	1964 - Open ROM								
0051239	Waspwood	55	S	0002196 Waspy	0081494	Travelog	56	M	0050252 Lowry Girl 114
0052102	Gray Chip	53	G	0009399 Sugar Tit	0083113	Quickwood	57	M	0010045 Bonita Adair
	1967 - Open ROM				0095050	Sierra Sugar	57	M	0077151 Sierra Candy
0052763	Lynwood	53	M	0009398 Miss Hondo	0095236	Big Three	57	G	0050252 Lowry Girl 114
0055344	Chipper Wood	55	G	0002689 Miss Hallie		1970 - Open ROM			
0055925	Just Seven	55	G	0008271 Keeta 7	0096403	Wooden Nugget	57	S	0008456 Nugget Hug
0055947	Calzona Katy	53	M	0011798 Smarty Pants	0096403	Blazey Wood	55	M	0028869 Bobbie Pin
0056543	Kim's Flirt	54	M	0003989 Sandy's Sister	0097884	Buttons Cumming	55	G	0050252 Lowry Girl 114
0056820	Hallie Wood	56	S	0002689 Sweet Hallie	0098735	Lucky Wood	57	M	0010772 Betsy Luck
	1966 - Open ROM								

Number	Name	Foaled	Sex	Dam No. & Name
0100756	Roanie Boy	53	G	0002194 Sage Hen
0111544	Little Speedy	51	G	0040458 Louise Schuyler
0113135	Mescal Brownie	55	G	0038245 Lazy Mark
	1960 - Open ROM			
	1960 - High Point Working Cowhorse			
0118074	Stormy Duke	55	G	T067035 Whoop De Do TB
0125210	Drifty Doll	57	M	U069219 Cantina
0132786	Driftoluck	56	G	0010974 Blue Berry
0174629	Tiny Tot Peake	54	M	U067783 Balmy Gal
0181281	Pine Top Peake	59	M	0009399 Sugar Tit
0181282	Polly Peake	60	M	0114070 Poppy Fields
0181283	Sassy Peake	60	M	0072983 Lil Troutman
0191211	Wander Wood	60	M	0063469 Bar Bee
0202117	Driftwood Monty	43	S	0002190 Hornet
0212559	Miss Driftwood	50	M	U078246 Pronto
	1958 - Open ROM			
0216047	Maggie McDrift	50	M	U067413 Adair mare
0321655	Zaca Mist	55	M	U079454 Shady Mazie
0478694	Runaway Peake	52	G	0005619 Shu Cat

APPENDIX 3

THE PRODUCE OF DRIFTWOOD'S DAUGHTERS

As broodmares, the Driftwood daughters foaled 583 AQHA-registered offspring. This carries on the family reputation of producing horses that "could do something under a saddle."

Dixie Belle National N003223
out of unknown mare
- Red Star Scimeca's

Unnamed mare by Driftwood
out of unknown mare
- Miss Billy Boy P
- Sleepy Miss P
- Miss Coolidge, *Racing ROM*
- Speedy Bill
- Miss Monte

Sally Wood
out of unknown mare
- Quitaque

Wanda Marie 3250
out of a daughter of Miller Boy
- Tan Jug
- Pancho Earl 2, *Open ROM*
- Senior Driftwood, *1966 Open ROM, High Point Steer Roping and High Point Steer Roping Stallion*
- Troubadour Dee
- Chubby Marie
- Goldy Binford, *show winner*
- Midnight Mist
- Senior's Nina
- Chester Jr.

Clayton Gal 3367
out of Red
- Ace
- Sallie Good'in
- Miss Nubbin
- Tater, *show winner*

Melody Boot 5299
out of Dolly
- Danny's Drift

Road Runner P 8848
out of Sage Hen
- Sum Son, *a prolific sire*
- Red Runner
- Finito, *Open ROM*
- Miss Pokette
- Road Show
- Real Cat
- Run Requested
- Miss Casbar
- Buddy Jim
- Jabali Squaw
- Ebb Tide, *Racing ROM*
- Miss Front Page

Stinger B 9174
out of Hornet
- Mr. Speedy
- Stinger Too Bars
- Mark Jr.

Nellie S 9363
out of Trixie
- Dolly M.
- Nellwood
- Wyno
- Miss Baby Snooks
- Flossy S.

Clara Bow M 9452
out of Trixie
- Jeeper Joe
- Waco
- Lady Bowie

Rosy D 9453
another daughter of Trixie
- Bessie B.

Blue Glory 10757
out of Straw
- Iron Bark
- Vaca Star
- Vaca Bluewood
- Glory Drawers
- Janita Smith
- Clabberwood
- Uphill
- Dan's Castle

Californiasweetheart 10758
out of Dos X
- Poco California
- California Honeygirl
- Big Jim Dandy
- Poco Gamble
- Red Bud Baby

Jabilina P 11959
out of Dos X
- El Bambino
- Roddy's 3 Bars
- Roddy's Vino, *show winner and Racing ROM*
- Roddy's Antonia, *Racing ROM*
- Jaba Black
- Penrod
- El Chongo, *show winner*

Katy Red 17278
out of High Tone
- Triple Chickmate, *show winner*
- Katy's Drift
- Red Roanoak

Fleet Wood 17337
out of Hightone
  Beaver Wood
  Dreamy Baby
  Little Spicy
  Beaver Fleet
  Lynn's Valey
  Quick Drift
  Merry Wood
  Gogi
  Fleet One

O See O 18660
out of Waspy
  Vee R Brownwood
  Deer Creek Dolly
  Osee Lizzy
  Miss Sun Dial, *show winner*
  Gatona
  Sea Green
  Miss Kittle
  Tom's Rosewood

Wood Wasp 19955
out of Waspy
  Miss Woodwasp, *show winner*
  Waspy Girl, *show winner*

Jima Red 20404
out of Jack Rabbit
  Berylewood
  Drifter Isle
  Barima Isle
  Jima's Choice
  Jima Wood

Baywood 21702
out of Coal Black
  Bay Leaf
  Dreamy Girl
  Spice Stick
  Brown Jacky
  Joe Wood
  Woodbox
  Baywood Dream

Driftalong Katy 22400
out of Freya
  Madera Roja
  Baby Sista
  Ket's Katy
  Ket's Drift Sox
  Breezealong Tony
  Katy's Luck
  Chacota
  RC Martin
  Redwood Dancer
  King Drifter
  Katie Chip, *show winner*
  Sessions Drifter

Buckwood 22531
out of Hancock Belle
  Bucky Wood
  Swap Wood
  Blaze Wood

Henny Penny Peake 22534
out of Sage Hen
  Jimmywood, *Open ROM*
  Henny Penny Poo
  Speedy Penny, *show winner*
  Silky Peake
  Red Pepper Peake

Two Pence 23197
out of Sister Penny
  Three Pence
  Pringle
  Miss Short Cut
  Runalyte
  Denny Dart
  Shingey Bay

Allen's Miss Skyla 24752
out of Skylark
  Drift Deck
  Spowan Barzell
  Drift Cloud
  Paul's Drifter, *show winner*
  Gotchen's Skylark
  Big Drifter
  Gotchen's Finessa

Sequoia Wood 25496
out of Gravy
  Driftwood Chant
  School Drifter
  Drifty Patch, *show winner*
  Driftwood Gold
  Driftwood Star
  Chip Wood
  Corkwood
  Cherrywood Gal

Miss Drifty 27477
out of Louise
  Drifty Barred
  Cocktail Susie, *stakes winner and Racing ROM*
  Super Drifty, *Triple AAA, Racing ROM and Superior Racing Top Eliminator*
  Drifty Steve, *show winner*

Mockey Bay 27977
out of Coal Black
  Fiesta Bay

Coal Bin 28563
out of Coal Black
  Noches Drifter, *ROM show winner*
  Possum Bin
  Bin Coal

Greasewood 31149
out of Hildegund
  Driftwood Dick
  Taliesen
  Dauchiwood
  Driftwood Scat
  Nobel Bar, *show winner*
  Greasebar, *show and race winner*
  Mi Caballo
  Vic Prince, *race winner and ROM*
  Alimony
  Dell Fury
  Rege Wood
  Drift Boots

Keetawood 31504
out of Keeta
  Poco Drifty
  Bar Bee
  Speedy Jane
  Poco Keeta
  Poco Mr. Chip
  Poco Rex
  Poco Bay Fern
  Poco Driftalona
  Driftbar Bob
  Mr. Keetawood
  Poco Keetawood

Miss Linwood 32525
out of Queen Ann
   Cibecue Roan, *ROM, Open High Point Steer Roping Stallion*
   Jake Stewart
   Pillo
   Times Tribute
   Vee R Foxy
   Vee R Ironwood
   One Nite Stand

Sierra Jetsam 32960
out of Sierra Buckskin Mary
   Doc's Drift Bar
   Stardrift Lady, *ROM*

Chilena 34066
out of Hancock Belle
   Redwood Beaver
   Stetson Bar
   Miss Chilena, *Racing ROM*
   Mr. Outside
   Frostens, *Open ROM and 9th World Champion Sr. Team Roping*
   Lena's Mark

Swiftwood 36318
out of Spiderette
   Podie Hancock
   Ravenwood
   Redwood Man, *ROM*
   Windwood
   Irishwood
   Peach Wood
   Mrs Fulton
   Deer Creek Juan
   Deer Creek Nancy
   Vee R Animoso
   Times Legend
   Driftwood Tonto
   Wooden Clown

Brown Beulah 38833
out of Queen Ann
   Vee R Peggy Ann
   Drifting May
   Wood Peake
   Beulah's Annie
   War Drift
   War Concho
   Mahala Hancock
   Woody Brown
   Medulce
   May Belline
   Redwood Jake, *show winner*

Gadget 41352
out of Louise Schuyler
   Go Wood
   Bud Parker
   Drift Reed
   Super Steel
   Trinket Que
   Miss K Gun
   Tio Red
   Small Town Girl

Travelog 81494
out of Lowry Girl 114
   Log Rhythm
   Miss Lowry Log
   Swap Cat
   Pali D' Or
   Log It
   Travel Talk
   Extricate
   Travellog's Johnny
   Take a Trip
   Cedarlog
   Wendy Went
   Tourista
   Red Log

Quickwood 83113
out of Bonita Adair
   Voo Doo Wood
   Arena Lady
   Go Quickly
   Twig D' Or

Sierra Sugar 95050
out of Sierra Candy
   Bates' Sugarwood
   Bates' Redwood
   Candy Wood
   Bates' Did Share
   Bates' Ironwood
   Bates' Duchess
   Sierra Taffy
   Bates' Reedetta
   Speedy Van Reed
   Bates' Ettareed
   Bates' Miss Pepper

Blazywood 96403
out of Bobbie Pin
   Miss High Drift
   Banner Blaze Babe
   Captains Lil Drift
   Joa Reed
   Reed Wood
   Blazy Bar
   Yankee Wood
   Danger Drift
   Miss Satinwood
   Snippywood
   Blazy Drift
   Oregon Sis

Lucky Wood 98735
out of Betsy Luck
   Lucky Drifter
   Lady Drift Reed
   Maplewood
   Lucky Speed
   Clovis Lucky
   Lucky Papi
   Miss Burgandy
   Peppy Pap

Drifty Doll 125210
out of Cantina
   Speedy Doll
   Driftless
   Money Player
   Musical Dolly
   Cholie Wood, *ROM show winner*
   Mickey's Dolly
   Mandy Wood
   Drifty Hill

Tiny Tot Peake 174629
out of Balmy Gal
   Catty Peake
   Plain Wood
   April Peake
   D' Or Drifty Doll
   Tuffi Bar Peake
   Willywood
   Luna Peake
   Wood Chex, *ROM and 1981 Open Superior Reining, Open High Point Reining and Open High Point Reining Stallion*
   Hol E Wood
   Brownie Driftwood

Pine Top Peake 181281
out of Sugar Tit
   Doc's Drifty Bar, *ROM*
   Drift Charge *1976 ROM, Open World Champion Sr. Reining and Open Superior Reining*
   Fair Drifter
   Fleety Peake

Polly Peake 181282
out of Poppy Fields
   Chicaro Polly
   Chuck Wood
   Coffee Flip
   Dunnie Wood
   Leslie Wood
   Odoms Pollywood
   Odoms Pollypeake

Sassy Peake 181283
out of Lil Troutman
   Lil Wood
   Sneaky Peake
   Missy Peake
   Run Little Sassy
   Bay Wood

Wander Wood 191211
out of Bar Bee
   Chillante Oro
   Our Par
   Miss Driftywood
   Driftwood Wanderer
   Top Wander Bid
   Last Drift
   Joe Jack Drifter
   Lady Cow Chip
   Molly Sullivan
   Blue Dee Fender
   Odoms Wanderwood
   Woodcharge

Chakaty 48005
out of Lady Lux N
   Artic Katy
   Toddy Wood
   Chatty Wood
   Chakaty Wood
   Katchina Wood
   Katy Blue
   Wooden Shiek

Cotton Cat 48556
out of Katy King
   Cat A Lac
   Perry's Cat, *ROM*
   Cottonwood Joe
   Sly Cat
   Cat Tod
   Johnny Cat
   Catch It
   Cotton Kitty
   Cat Wood
   Piojo
   Baywood Cat, *ROM*
   Bras Cat
   Cotton Belle
   Bay Whisk

Wood Start 50017
out of Beater
   Top Start

Woodkee 41510
out of Keeta 7
   Woodkee's Joker
   Drifting Dart
   English Leather
   Colectondelivery

Driftalena 42198
out of Lady Lux N
   Bunny's Dart
   Driftabar
   Anglewood, *show winner*
   Bonnie Tu Belle
   Miss Chinaco

Miss Wood 43632
out of Miss Limit
   Wood Cock, *show winner*
   Lowl's Drifter, *show winner*
   Bates' Stormy, *show winner*
   Bates' Drifter
   Bates' Chipette
   Poco's Pleasure
   Bates' Firewood

Quita Bonita 44578
out of Sweetwater Sue
   Pancho Drift
   War Tax
   Mojave War

Woodfern 47646
out of Keeta 7
   Oui Oui
   Poco Fernwood
   Drifter's Peppy
   Poco Drifty
   Mr. Driftwood
   Poco Drifty Girl
   Poco Whirlwind
   Poco Whizwood
   Poco Driftaleta
   Rawhide Poco
   Speedy Lady
   Wagwood
   Wildwood Miss
   Poco Chileta
   Drifty's Poco

Judy Sue 80515
out of Nita M
   Poco Katy Peake
   Poco Judy Sue
   Poco Dolletta
   Poco Hallywood, *show winner*
   Speedy Judy
   Judywood
   Another Judy Sue
   Poco Skywood
   Miss Judy Sue
   Mr Jetwood
   Poco Scatcat
   Jo Jo Wood
   Scottwood
   Poco Wood Chip
   Poco Firewood
   Poco Firebrand, *show winner, ROM*

Tweedle Dum 81493
out of Lucky Lady Tucker
   Chips D' Or
   El D' Or, *show winner*
   Tweedle D' Ora, *ROM*
   De Dum D' Or
   Bras-wood
   Accomodator
   Hum Dum
   Dum Walter
   Blazing Blanton
   Lucky Lady Win
   Lucky Cal
   Indirect
   My Boss Rich
   Direct Drift
   Spot's Peake
   Poly Bob
   Bobby's Hit

Bobbinet Peake 57828
out of Bobbie Pin
  Spect's Peake
  Bobby Twaine
  Poly Bob
  Bobby's Hit
  Yeller Lust
  College Kid
  Rob Bob
  Bob D'Or
  Bobbin' Or

Filet Mignon 58912
out of Jamboree
  Spinaway of Laguna
  Granymede
  Spray of Laguna

Oakwood 62027
out of Oklahoma Gal
  Miss Drift
  Drifter D'Or
  Oakwood Amy
  Oakwood Annie
  On'ry Cuss
  Mr Oakwood, *show winner*
  Poco Woody Drifter
  Poco Rajah

Tweedle Dee 62168
out of Tuckaluck
  Tweedle Bar, *show winner*
  Top Deduction
  Tweedle Tuck
  Snap Crackle Dee
  Star Dee Barred
  Docs Lucky Dee
  Miss Tweedle Bar
  Lucky Dee Girl
  Tweedee
  Dee River
  Docs Dee Dee Bar
  Mr Tuck Away
  Be Somethin'
  Color Time
  Solid Tweed
  Win Petite Dee
  Tweedle Tim

Bugaboo Peake 63070
out of Honey Bug
  Film Folly
  Drifter Uno

Quick A Lick 63265
out of Nugget Hug
  Direct Line
  Poly Bill
  Quick Chant
  Hit and Win, *show winner*
  Quick A Long, *show winner*
  Roan Licker
  Bit A Quick
  Warwood
  Hit A Lick
  Wonder Lick

Kitty Wood 63470
out of Shu Cat
  Prontotito
  Shu Kitty
  Kitty Colt One
  Cat Talk
  Win Cat
  Oso Lucky, *ROM in Racing and ROM in Youth Performance*
  Billy Dilly
  Naughty Wood
  Fire Bug Babe

Annie Wood 63577
out of Queen Ann
  Ante Over
  Amigo Red Boy
  Lady Wood
  Teak D'Or
  Doc's Anna Wood
  Orphan Andy Wood
  Annie Driftwood
  Doc's Drifter, *show and NCHA winner*

Wood Mite 63993
out of Spiderette
  Betsy Chant
  Catty Mite
  Mite Bar
  Driftwood Kitty
  Bobby Mite
  Driftwood Dog
  Drifty King Chex
  Mite Wood
  Drifty Fritzi Chex
  Wood Chipper

Misty Miss 64272
out of Reed's Hannah Hancock
  General Pickett
  Vee R Baywood
  Plenty Wood
  Deer Creek Ella, *show winner*
  Hard Times Inice

Armell 64503
out of Sugar Tit
  Trouble Wood
  Miss D'Or
  Wood Bee
  Tuffwood
  Chief Charley
  Wood Chief
  Georgia D'Or
  Willie Wood
  Terry Tivio
  Darmell
  Rural Wood

Rosewood 66244
out of Miss Hondo
  Double Drift, *ROM and prolific show winner*

WPR Lucky Drift 69686
out of Miss It
  Drift Chick, *Open ROM, Youth ROM, and show winner*
  Drifton, *show winner*
  Driftette
  Drift A Go Go
  Mr. Wood Sen
  River Wood
  Elegant Chex
  My Lucky Hobby
  Bejac Chex

Zoe Driftwood 74813
out of Dos X
  Joe's Orphan Ann
  Joe Driftwood

Sierra Redwood 76443
out of Sierra Roxie
  Sierra Sequoia, *show winner*
  Sierra Speeder
  Yens Driftwood
  Roxie Drifter
  Driftwoods Roxie

Tooka Peake 77753
out of Lucky Pat
   Sneak A Peake
   Breeze Wood
   Elden, *show winner*
   Rojo Cielo
   Baronesa

Wood Mist 78206
out of Dusky Ruth
   Misty Knox
   Bresa Mist
   Wood Mist Gold
   Wood You Bar, *ROM and show winner*
   Woody Peake
   Fortuna Drift

Drift Girl 78235
out of Dr. Charlotte
   Radel King
   Mid Drift
   Snip Drift Girl
   Whoa Man Booger, *show winner*
   Drift Girl Tivio
   Pretty Drifter
   Super Drift, *show winner*
   Nugs Drifter
   Pocos Drift Girl
   Red Bar Drifter

Speedy Hug 79726
out of Wanda C
   Easterwood
   Speedy Barred

Calzona Katy 55947
out of Smarty Pants
   Clover Pepsi
   Clover Drift
   Clover Roo
   Katy's Image
   Velvet Drift Clover Cherry
   Clover Woodie, *show winner*
   Clover Gidget

Kim's Flirt 56543
out of Sandy's Sister
   Tecolote Tivio
   Speedy's Sweetie
   Taco Tivio
   Johnny Kim
   Dusky Kim

Hug Me Tight 56821
out of Nugget Hug
   Barred Hug
   Super Tight
   Drifty Bar Hug
   Hug Me Chick, *2nd NCHA show winner*
   Hug Me Green
   Tight Secret

Calzona Babe 57009
out of Smarty Pants
   Miss Ikewood, *show winner*
   Yum Yum Wood
   Frosty Special
   Win Drifter
   Willy Maykit

Wood Start 50017
out of Beater
   Top Start
   Beater Bar
   Kenito
   Drifty Blue Bars

Speedy Scat 50593
out of Shu Cat
   Oops Cat, *show winner*
   Speedy Anson
   Miss Stand Pat, *ROM*
   Worry Scat
   Sugar Scat
   Hobby Scat
   Scat Ballou
   Sure Scat, *show winner*
   Speedy Bingo
   She Scat

Lynwood 52763
out of Miss Hondo
   Poco Drifter, *ROM*
   Mr Hotdogger
   Bobby's Drifter
   Tuny's Bea Haven
   Tuny's Drifter
   Otto Bar's Pard
   Hoby Beck
   Wormwood, *show winner*
   Dublin's Drift
   Dublin's Miss Wood

Maggie McDrift 216047
out of an Adair mare
   Driftwood Roper
   Jack the Roper

THE AUTHORS

Photo by Jim Crocker.

Photo by Livingston.

Phil Livingston—Weatherford, Texas

A "horse-crazy" kid who grew up to rope, rodeo and ranch cowboy, Livingston fell in love with the Driftwoods during the middle 1950s when he attended Cal Poly at San Luis Obispo, California. "Too many of the tough ropers were riding the Driftwoods not to be impressed with them," he remembers.

In 1962 Livingston went to work for the *Western Livestock Journal,* an important step since he began to write on horse subjects there. He also, on a visit to the Gill Ranch, met Jim Morris, an event which forty years later would result in this book. The *Livestock Journal* was followed by a ten-year stint as Advertising Manager for Tex Tan Western Leather Company in south Texas, the same position at the well-known Ryon's Saddlery in Fort Worth, and 3 years as Editor of the *Paint Horse Journal.*

As a writer his by-line has appeared on several hundred articles which have appeared in most of today's horse periodicals. Livingston has penned numerous biographies on the early Quarter Horse stallions over the years, concentrating on those who produced good usin' horses. He has also authored several books, including the definitive work on the United States Remount Service and its contribution to America's light horses, *Team Penning* published by *The Western Horseman,* and contributed to the

Jim Morris—Exeter, California

A native Californian, Jim Morris grew up around horses. His father was a stock farmer in Exeter, a town nestled next to the Sierra Nevada mountain range, and used the animals extensively. From childhood Morris was exposed to the animals, both saddle and draft types, and learned to appreciate conformation, athletic ability and a willing mind. His first venture into the breeding business occurred when he was 13. He bred his saddle mare to a nearby Thoroughbred Remount stallion. The fact that he lost the colt as a yearling didn't deter him from making a lifelong career in the horse business.

Except for a three-and-a-half-year stint flying 52 missions during World War II, Jim Morris has earned his livelihood in some form of agriculture. He's farmed, cowboyed, run steers, managed citrus groves and owned a saddle shop in nearby Visalia. During the late 1940s and '50s he was aware of the impact Driftwood was making on the using horses of the Pacific Coast, visited Rancho Jabali several times, and decided to get back into the breeding business. He remembered seeing Driftwood and Speedy Peake and was impressed with the horses.

His first stallion was Old Granddad, a full brother to the well-known Poco Bueno. He had been impressed with the King-bred mares which Rancho Jabali and

Phil Livingston

Western Horseman Legends series. It was his biography on Driftwood in *Legends II* which helped spark the idea for a complete book on the bay stallion.

In addition to his advertising and writing commitments, Livingston continued to rope on a part-time basis. He has always been interested in the breeding behind the top horses of the sport and discovered that certain Quarter Horse families seem to proliferate in the arena. As a roper, Phil Livingston is a strong believer in the old saying, "You gotta be mounted to win." With the Driftwoods, a roper is.

Jim Morris

Perry Cotton were using and the foals which they were producing. In the 1960s, faced with juggling his business commitments and hauling his daughter Michele (she qualified for the AQHA Junior World Show in the Stock Horse division) to shows, Morris sold the stallion and small band of mares.

Shortly after retirement in 1987, Jim Morris visited Russell Keely, saw the week-old Lindsay Peake, and had to have him. The good looking brown stud colt combined the modern bloodlines of Bar Girl (Son O Sugar/Doc's Bar Girl) with the proven performance of Driftwood down through Speedy Peake. He hunted down five granddaughters of Old Granddad to form his mare band. Two daughters of Speedy Man, Lotta Driftwood and Speedy Lil, were leased from Henry Kibler. Morris eventually purchased Lotta Driftwood. When crossed with Lindsay Peake, the foals carried some of the most intensive Driftwood breeding in the country.

Since then, Jim Morris has become deeply involved with the Driftwood tradition. He was one of the founders of the informal Driftwood Breeders Association, put out a quarterly Driftwood Breeders Newsletter, and helped develop the network between bloodline enthusiasts.

A visit to Texas in 1999 rekindled an old friendship with Phil Livingston. A second visit, a year later, much discussion over a strain of horses that both men loved and respected, and the decision was made to research, write, and publish this book. Without Jim Morris' contacts within the industry, his extensive knowledge of bloodlines, and especially his enthusiasm, the project would never have gotten beyond the talking stage.

BIBLIOGRAPHY

Porter, Willard H. *13 Flat*, A.S. Barnes and Co., Inc., Cranbury, New Jersey - 1967

Porter, Willard H. "Driftwood—One of the Best Rope Horse Sires in the World is Dead", *Hoofs and Horns* - December 1960

Livingston, Phil. "Driftwood", *Legends II*, published by *Western Horseman Magazine*, Colorado Springs, Colorado - 1994

Thornton, Larry. "The Story of Driftwood", *The Southern Horseman*, September 1997

Thornton, Larry. "Driftwood—The Early Years"; *Foundation Quarter Horse*, Sterling, Colorado - 1998

Peake, Katy. "Tribute to the Rope Horse"; *Quarter Horse Journal*, Amarillo, Texas - April 1953

Simmons, Dianne. *A Faded Past*, unpublished manuscript - about 1993

Morris, Jim. "The Driftwood Legend Continues"; *Driftwood Breeders Newsletter*, Volume 1, No. 2 - January 20, 1996

Nicholson, Barbara Turner, "Turner Family", *Driftwood Breeders Newsletter* - July 20, 1997

Wright, Brian. "Driftwood Remembered", *The Remuda*, Talpa, Texas - 1998

Kenny, James. Telephone interview with Phil Livingston - October 11, 2001

Seals, Nita. *Ponder—The Little Town With the Big Rodeo*; privately printed - 1985

Porter, Willard H. "Roping Horse Sires"; *The Quarter Horse Journal*, Amarillo, Texas - April 1953

Brisco, Jack; Ponder, Texas. Interview with Phil Livingston - Febuary 3, 2001

Seals, George; Ponder, Texas. Interview with Phil Livingston - Febuary 3, 2001

Webster, Catherine Peake. "Down Memory Lane"; *Driftwood Breeders Newsletter*, Volume 1, No. 3 - July 1996

Webster, Catherine Peake. "Down Memory Lane", *Driftwood Breeders Newsletter*, Volume 2, No. 1 - October 20, 1996

Wright, Brian. *The Driftwood Directory*; published by DBA2 Publishing, Arlington Heights, Illinois - 1997

Morris, Jim. "Sale—No Sale"; *Foundation Quarter Horse Journal*, Eugene, Oregon - April 1998

Morris, Jim. "Speedy Peake"; *Driftwood Breeders Newsletter*, Volume 1, No. 4 - September 1, 1996

Morris, Jim. Numerous conversations on his memories of the Driftwood horses and the people who owned and rode them

Cotton, Perry and Evetts, "Hoke". Interview with Jim Morris - October 30, 2000

Haskell, Melville H. "Racing Quarter Horses"; *Southern Arizona Horse Breeders Association*, Tucson, Arizona - 1943, 1944, 1945

Porter, Willard H. "Driftwood Blood Still Popular"; *The American Horseman Journal* - December 1981

Sheppard, Chuck. Statement from Rancho Jabali - March 29, 1955

Rockingham, Montague. "Old Tony—the Wear Horse"; *Quarter Horse Journal*, Amarillo, Texas - April 1952

Porter, Willard H. "Four-Legged Cowboy"; *Hoofs and Horns*, Tucson, Arizona - October 1950

Webster, Catherine Peake. "Down Memory Lane"; *Driftwood Breeders Newsletter*, Exeter, California - September 1, 1996

Morris, Jim. "Roy Wales"; *The Driftwood Breeders Newsletter*, Exeter, California; Volume 2, No. 3 - April 20, 1997

Porter, Willard H. "Chip's Still in the Game"; *Quarter Horse Journal*, Amarillo, Texas - August 1964

Smith, Grace. "Pat's Roany"; *Hoofs and Horns*, Phoenix, Arizona - July 1960

Warden, Theresa Ann. "Driftwood Ike—Rope Horse Legend"; *Buckskin Horse Magazine,*

AQHA Get of Sire Performance Foals - Chipperwood 55,344

Morris, Jim; Exeter, California. Notes of Chipperwood foals' performance

Morris, Jim. "Double Drift"; *The Driftwood Breeders Newsletter,*
Exeter, California, Volume 2, No. 1 - October 20, 1996

Warden, Theresa Ann. "Driftwood"; *Foundation Quarter Horse Magazine,* Waldron, Arkansas - January/Febuary 2001

The Driftwood Blood Bank. Advertisement for the Potter Ranch, Mariana, Arizona; *The Foundation Quarter Horse Journal,* Eugene, Oregon - February 2000

Livingston, Phil. "The Driftwood Influence"; *The Driftwood Horseman,* DBA2 Publishing, Arlington Heights, Ill. - 1997

Williams, George. "Dale Smith—Rodeo's Fearless Leader"; *Persimmon Hill,* published by the National Cowboy Hall of Fame, Oklahoma City, Oklahoma, Volume 6, Number 2 - 1976.

Rancho Jabali breeding and foaling records, business statements, listing of outside mare owners and horse purchases, correspondence, etc. - 1942 through 1963

"Quarter Horse Breeding Program Started at Cal Poly"; *Santa Barbara News-Press,* Santa Barbara, California - October 18, 1955

Cotton-Peake Sale Catalog - November 14, 1948

Cotton-Peake Sale Catalog - November 12, 1950

Rancho Jabali Dispersal Sale Catalog - August 14, 1966

Advertisement for Quicksand - Cotton-Peake Quarter Horses; *Quarter Horse Journal,* Amarillo, Texas - April 1949

AQHA Get of Sire List for Driftwood - 1988

Barnett, Jo-ann Rosser. "Cow Horse Hall of Fame"; *Working Cow Horse Breeders of Yuba-Sutter Counties, Inc;* Yuba City, California - 1984

Monnens, Tony. "A Tribute to My Friend and a Great Horseman—Stanley Johnson"; *The Driftwood Horseman,* DBA2 Publishing, Arlington Heights, Ilinois - 1996

Warden, Theresa Ann. "Lone Drifter, Mel Potter . . . and family"; *Buckskin Horse Magazine,* page 9

AQHA Dam's Sire Record (Driftwood) - 1983

AQHA Get of Sire List for Orphan Drift - 1987

Haythorn Ranch 1998 Production Sale Catalog, Arthur, Nebraska - 1998

Smith, Dale with Porter, Willard. "Team Roping Horses Have to be Tough"; *The Western Horseman,* Colorado Springs, Colorado - July 1971

Tumlinson, Monroe; Cresson, Texas. Conversation with Phil Livingston - November 3, 2000

Rodeo Sports News - 1975 Championship Edition

Thorson, Freeland; Nampa, Idaho. telephone conversation with Phil Livingston - November 5, 2000

Thorson, Freeland; Nampa, Idaho. "The Story of Double Drift"; *The Foundation Quarter Horse Journal,* Eugene, Oregon - April 1998

Hall, Cheryl; Paradise Valley, Nevada. "Double Drift Recollections"; *The Foundation Quarter Horse Journal,* Eugene, Oregon - April 1998

Ginsburg, Debra. "Perry and Eunice Cotton: Six Decades in the Company of Horses"; *California Thoroughbred* - June 2000

Morris, Jim. "War Drift and Tex Oliver"; *Driftwood Breeders Newsletter,* Exeter, California, Volume 2, No. 2 - January 20, 1997

Montgomery, Carleigh. "Hug Me Chick"; *Quarter Horse News,* Fort Worth, Texas - October 25, 1983

Miller, Christie. "The Eliasons: Driftwoods in the Dakotas"; *The Foundation Quarter Horse,* Sterling, Colorado - October-November 2000

Will Gill & Sons Ranch biography, supplied to Jim Morris by David Gill - November, 2000

Porter, Willard H. Letter to Katy Peake from Cowboy Hall of Fame inviting her to Driftwood's induction, and the placement of a plaque, on the Trail of Great Cow Ponies - October 21, 1983

Balkenberry, Sammy Fancher. Telephone conversation with Jim Morris - December 2, 2000

Balkenberry, Sammy Fancher. Taped interview with Jim Morris - January 8, 2001

McCraine, Kathy. "Babbett Ranches"; *Foundation Quarter Horse Journal*, Eugene, Oregon - October 1999

Stallion Listing - 2000 Babbett Ranch Sale Catalog

Clanahan, Holly. " 'Drift'ing Tradition - A Story of the Will Gill & Sons Horses"; *America's Horse*, published by the American Quarter Horse Association - January/Febuary 2001

Arnold, Perry. Telephone interview with Phil Livingston - March 14, 2001

Pixley, Arlene. 4P Cattle Company information and biography to Jim Morris - January 29, 2001

Sterns, Rhonda Sedgwick. "Bigger . . . definitely better"; *Tri-State Livestock News Horse Edition*, Sturgis, South Dakota - January 2001

Fincher, John; Plattsmouth, Nebraska. Telephone interview with Phil Livingston - March 27, 2001

Moench, Gene; Valparaiso, Indiana. Telephone interview with Phil Livingston - April 24, 2001

Morris, Jim; Exeter, California. Numerous discussions on the formation of the Driftwood Breeeders Association with Phil Livingston

Wright, Brian; Arlington Heights, Illinois. Discussions with Phil Livingston regarding the formation of the Driftwood Breeders Association

Driftwood Breeders Newsletter, Exeter, California, Volume 1, No. 1 - October 20, 1995

Potter, Mel; Marana, Arizona. Telephone interview with Phil Livingston - March 18, 2002

Burson, Johnny; Silverton, Texas. Telephone interview with Phil Livingston - May 1, 2002.

Denhardt, Robert M. "Foundation Sires of the American Quarter Horse"; University of Oklahoma Press - 1976

www.ingramcontent.com/pod-product-compliance
Lightning Source LLC
Chambersburg PA
CBHW080414170426
43194CB00015B/2803